Transactional Lean: Preparing for the Digitalization Era

Bruno G. Rüttimann

Transactional Lean: Preparing for the Digitalization Era

A Systematic Approach to Industrialize Office Processes

 Springer

Bruno G. Rüttimann
D-MAVT - IWF
inspire AG/ETH Zürich
Zürich, Switzerland

ISBN 978-3-030-22862-0   ISBN 978-3-030-22860-6   (eBook)
https://doi.org/10.1007/978-3-030-22860-6

This Springer imprint is published by the registered company Springer Nature Switzerland AG
The registered company address is: Gewerbestrasse 11, 6330 Cham, Switzerland

Sapientia per actum bonum

Foreword

There is hardly a company in high-salary countries that is not being forced to optimize its production in order to reduce cost, meet delivery times and reach quality. Therefore, many of these have applied the Toyota Production System in production. Some renamed it or developed similar ones to reach acceptance within their companies. Lean Management, as it is called by Womack and Jones, also influenced other branches such as IT and Health.

As production contributes less and less to the total cost of a product, other parts of the value chain to be optimized come into focus. This can be seen in the appearance of new terms such as Lean development, Lean procurement or Lean accounting. These processes have an administrative character and take place in offices, so Lean Administration or Lean Office might be used as the overall concept.

The application of Lean principles to the office environment is still in its early stages—but it faces difficulties and has not led to similar success as in production so far. Most of the effort to lean out offices are currently only being focused on 5S activities such as workplace and file readiness. Clean and orderly workplaces, cleansed data and common folder names are valuable outcomes, but Lean Management offers more.

When looking at literature and research, no serious effort was made to transfer the Toyota Production System to the administrative processes, except for some hands-on "how to do" books. Based on his in-depth work "Lean Compendium" Bruno Rüttimann develops a consistent theory of Lean for the administrative processes that go beyond pin pointing some aspects. This book is the first of its kind. His profound mathematical knowledge enables Rüttimann to formalize the Lean principles and to make them applicable for further research. His long experience as a practitioner makes the book a valuable source for practice.

Kufstein, Austria

Prof. (FH) DI Dr. Martin Adam
Director of Bachelor and Master Degrees Programs
Professor for Lean ERP Information Systems
and Business Process Modelling
University of Applied Sciences FH

Acknowledgements

My very special thank goes to Urs Fischer as well as Antonio Gallicchio. Both have not only contributed by giving valuable comments to the content but also by writing three sections of this book enriching the content.

Thanks also to Dr. Martin Stöckli, responsible of the inspire academy as well as COO of inspire AG, a technology transfer institute of ETH Zürich, for supporting the present book project and adopting the main content as basis for the inspire Lean Office training courses.

Prologue

The Toyota Production System (TPS) has changed the way that products are manufactured. The TPS has been a revolutionary production system not only for the automotive industry. Today hardly a single industry is not implementing Muda-free, or better, a Kaizen-based JIT production. The benefit of Lean to reduce waste in office processes has conquered also transactional service companies, however with contingent success. Indeed, the TPS has been described in several books and has become better known in the Western world with the word Lean reducing Muda. Although it is true that by applying the TPS techniques Muda will be eliminated, it has to be pinpointed, that by applying some Muda-reducing techniques, such as e.g. 5S, Lean value-add based process analysis, one is not implementing forcedly the TPS Lean system, although often believed so. In fact, the TPS is a comprehensive production system going beyond the application of some tools. Nevertheless, through the high potential of cost savings and the aim to becoming more competitive, Lean has conquered the taste of executive management leading to the introduction of so-called operational excellence initiatives (OPEX) in many industries. Therefore, Lean has also been introduced in transactions-based service companies, such as credit institutes, insurance companies, telecom, or hospitals and many other institutions. Under the label of Lean management, the ideas of the TPS have found their applications also in the office environment and business transformation in general.

Whereas I can live in the context of an office environment with the widespread notion of Lean as waste reduction based on "discovering Muda by Gemba walk", I prefer to talk within an industrial environment about a "Kaizen-based JIT production"—identifying the TPS only with Muda (waste) reduction is really by far too restrictive. Indeed, the TPS embodies a new production theory based on cellular manufacturing with demand-pull single piece flow (SPF) allowing nearly stockless just-in-time (JIT) manufacturing. This way to produce is very different from classic Western batch & queue (B&Q) push-production philosophy and is also applicable "beyond large scale production" of Ford's assembly lines as Taiichi Ohno told. The superiority of the Toyota paradigm based production system is incontestable.

Although several basic Lean tools have been introduced to the office world, such as value stream mapping (VSM), 5S workplace organization, value-add and non-value add concept, zero error culture of first time right, different than in the manufacturing industries, generally, the OPEX initiative had only limited success. The reasons are multiple, as we will see. At this point, the question arises: is it even possible to apply a TPS-derived Lean to transactions based service industries? Frankly speaking, according to my modest experience in 20 years of management background and then afterwards as consultant as well as lecturer, rarely I have seen a consequent implementation of the TPS in an office environment. Different reasons stay at the base, but are mostly linked to the wrong techniques applied to the service industry or the office world, having a different intrinsic way of transforming input into output. I believe that the limited concept of interpreting Lean as waste reduction is the main cause of failures. Interpreting Lean as a new way to work beyond Muda and Kaizen will open an additional potential in gaining competitiveness also in the office.

However, an additional challenge is turning up: the era of new digitalization, which will change the way we work. Deploying today OPEX is absolutely mandatory; not yet being lean might compromise the future of the company. Indeed, the final aim is not OPEX but BEX (business excellence). Ongoing digitalization and progress of artificial intelligence (AI) will change today's business models. For this challenge, Lean is only the foreword. In that sense, we have to interpret the words of UBS CEO Sergio Ermotti "we have to industrialize the bank system". Therefore, in this book, we will present a new industry-paradigmatic TPS approach to the office world to exploit the whole potential of Lean and preparing processes as well as the way of value-generation to master the new digitalization challenge.

Going beyond Muda and Kaizen in the office means to apply an industry-derived approach to the transformation of how to work in transactions-based offices. The success will be based on adopting the whole TPS derived toolset; indeed the TPS tools are neither a toolbox from which to choose just some tools nor, even less, a mindset as often improperly divulged by consultants—however, they are a comprehensive synergic tool system as we will see. This book will not replace existing books, where many TPS tools are presented for the office world; this book complements existing Lean Office literature with the remaining additional toolset opening a new perspective on transforming efficiently and effectively inputs into outputs. However, we will not enter in basics, so this book is rather destined to a knowledgeable readership. Each section of the central tools chapter will present the original idea behind each TPS tool applied in the industry context; indeed, one has to have understood the tools in the original context for which they have been developed. Then, the single tool is transposed and adapted to the special characteristic of the transactional context. Each tool section of this vademecum will have an easy to remember implementation principle, which summarizes the guideline to follow for a successful Lean office transformation.

In the following, we will generally talk about Lean Office, comprising also the topic of Lean Administration. Indeed, with Lean Office is usually understood the application within service industries whereas with Lean Administration is usually intended the administration processes of a manufacturing industry. The here presented concepts are universally valid in transaction-based environment.

Realizing the envisaged profound changes to the actually applied Lean office waste-reduction approaches, the title of this book could also have been "Beyond Waste Reduction in Office Processes" stressing the industry-paradigmatic approach of Taiichi Ohno's book; nevertheless, I preferred to stay with a more sober title giving a clear message of content. Indeed, the content of this book has a revolutionary character for office application and reveals a new approach to boost productivity for transaction-based processes preparing for the digitalization era. However, the challenge will not be the application of the adapted tools, but to overcome the potential resistance of employees to change their present rather comfortable office working habits. In any case, be confident in your actions and enjoy first reading the book by gaining new insights regarding Lean applied to office.

Zürich, Switzerland   Dr.-Ing. Bruno G. Rüttimann

Contents

About the Author

Dr.-Ing. Bruno G. Rüttimann, MBA studied electronic engineering at the Polytechnic Institute of Milan and business administration at the Bocconi University. During 20 years of an international management career in multinational companies, he was appointed for different positions such as managing director, director of business development, strategic planning, or continuous improvement, gaining deep insight in management, M&A, industry logics, globalization, as well as operational excellence. He is a well-known speaker at international industry and academic congresses having published several academic papers and articles with topics of globalization and lean manufacturing. His main contribution consists of having developed new global trade and FDI models as well as formalizing Lean into a consistent manufacturing theory. Now, he is Lecturer at ETH Zurich as well as consultant at inspire AG, a technology transfer institute of ETHZ.

Chapter 1
Introduction

The people's mindset of values reflects the way of thinking and acting. The mindset of Western managers is quite different from that of Japanese managers. This derives from cultural background and has also been a direct consequence of scholastic education. Let us have a look at two examples, the first one how to approach problems, and the second one how to prioritize targets.

Modeling of production systems is a sophisticated task because the optimization of such complex systems is not trivial. Different from physics or mathematics, which have their own deterministic laws, production theory consist of the allocation of scarce resources in a non-deterministic environment satisfying economic laws of profit maximization. Indeed, the difficulty to solve e.g. a production-scheduling plan is related to optimize a target function in concomitance of a restricted system of resources. However, instead of solving complex equation systems, maybe even of NP-type of problem complexity, one could also change the problem setting. Indeed, at the same time as Western production engineers have put great effort to optimize scheduling of manufacturing systems, the original Toyota production system (TPS) has been conceived avoiding scheduling issues; a complete different, diametral-opposed approach to solve problems.

Let us have a closer look to that in order to better understanding the complexity without stressing to much the associated math. The optimization problem may have different targets such as minimizing manufacturing costs, maximizing margin contribution, or reducing process lead-time (PLT), always exploiting best existing production resources. Such a linear optimization exercise can be described by an objective function to be optimized and a set of restrictions. The matrix $\mathbf{A}$ of restrictions $\mathbf{Ax} \leq \mathbf{b}$, where $\mathbf{A}$ is $[a_{mk}]$, denotes the unitary specific absorption of the machine m by the product k. The vector $\mathbf{b}$ denotes the absolute capacity, and $\mathbf{x}$ is the vector of the product mix with $\mathbf{x}^*$ denoting the solution vector of the optimal mix to maximize the objective function $z = \mathbf{c}^{\mathrm{T}}\mathbf{x}$, where the transposed vector $\mathbf{c}$ denotes the specific margin contribution vector. First applications of linear optimization were made in production planning which gave the name of Linear Programming (LP) to this new branch of mathematics called Operations Research. The LP systems of type

© Springer Nature Switzerland AG 2019
B. G. Rüttimann, *Transactional Lean: Preparing for the Digitalization Era*,
https://doi.org/10.1007/978-3-030-22860-6_1

$$\begin{cases} \max\{z\} = \max\{c^T x\} \\ A \cdot x \leq b \\ x \geq 0 \end{cases}$$

allows to maximize a given production system but are not especially apt to model Lean type of manufacturing systems. Do not fear, we will limit here our excursion in mathematics and refer to the paper [1] for interested readers. However, the difference between LP and Lean Manufacturing (LM) is profound and shows once more the difference of Western and Japanese thinking. Western engineers try to optimize a static system modeled by an equation system, which can be called ex-ante optimization leading to an optimized production scheduling to be observed; MRP and ERP (manufacturing/enterprise resource planning) systems have followed. On the other side of the Pacific Ocean, Toyota did not plan to develop a sophisticated system of production scheduling—no, Toyota changed the problem setting simplifying it by developing a new way to produce: from push to pull. Toyota's Lean production system reacts real-time to change on demand and produces just-in-time (JIT) exactly what has been consumed without planning necessity. At the base of JIT stands a single piece flow (SPF) which can be described with

$$SPF := \left\{ \lim_{n \to 1} Push\{B_k(n)\} = \lim_{n \to 1} Pull\{B_k(n)\} | CT_i = CT_{i+1} \right\}$$

and by introducing this type of transfer principle combined with the Kanban system, production became apt also for "beyond large scale production", as Taiichi Ohno postulated [2]. With that, Toyota revolutionized manufacturing not necessitating to planning and optimizing production scheduling. What for a different approach to solve production problems!

The second example reflects the value system and what to put at the top of the list from the target function. Management problems are also production-related problems; indeed, the transformation of input factors into output stands at the base of every operation. If we give managers three items to be ranked according to their believed importance, such as

- sales which can be assimilated to throughput,
- cost which can be assimilated to headcount, and
- net working capital (NWC) which can be assimilated to work-in-process (WIP),

Western managers come to another ranking than Japanese managers. The European manager would probably rank highest:

1. Cost
2. Sales
3. NWC

and the Japanese manager would probably rank:

1. Sales
2. NWC
3. Cost.

This resulting ranking is very different and could be interpreted with the degree of knowledge about the TPS. The cost-orientation is typical of Western management; one could also argue this is even further driven by explicitly reducing waste (Muda) which bear non-value add and therefore cost. However, the Japanese manager's mantra is based on throughput, which reflects customer's imposed takt rate, low WIP improves not only financial liquidity but also speeds-up production lead time assuring on-time-delivery, and finally, cost are a resulting aspect of efficient and effective management. Again, what a different outcome is resulting from the evaluation of the same situation.

We could even go further and state that the usually applied and commonly shared specific margin contribution logic, i.e. margin per piece, has become obsolete. Indeed, today the high performance industry logic has switched to margin contribution per hour logic, which by the way has the throughput logic at its base.

These examples are given because office processes are neither LP-optimized nor LM-organized. However, they can be conducted to similar principles applied in manufacturing, as we will see. Office processes bear the highest waste with white-collar employees being higher paid than blue-collar workers and show a very high potential for improvement. Therefore, the waste reduction approach had been at the center of OPEX initiatives in Europe (and the US). First shy approaches to reduce Muda (waste) in transactional environments have been made but with limited success. We need to change our explicit cost-based Muda-paradigm adopting a different way to approach challenges.

With this book, we will go beyond the "traditional" Lean Office approach of searching Muda by Gemba walking the offices, often represented in consulting presentations by a small man holding a big lens in his hand. However, we will apply an industry-derived comprehensive approach to industrialize office processes to make them lean, i.e. efficient and effective. Off course, Muda may also be detected by Gemba walk as already Taiichi Ohno told us for the industrial environment [2], however, Muda is not always immediately visible in the office! The concept of Gemba walk (Gemba means literally: place where something happens) is of essential concept in Lean manufacturing; one has to go oneself to look where people are working to build his own undistorted opinion about the situation. Indeed, in Western management culture, it is usually the assistant to the CEO going to shopfloor to see the "crime scene" and then reporting to him; the CEO might be too busy for that. In the Japanese management culture however, it is the CEO personally going to the Gemba for making his own undistorted opinion to evaluate better the present situation.

Nevertheless, Lean in the office environment seldom has gone beyond VSM (Value Stream Mapping), which by the way has often been simply mapped as a process diagram, better known with the word of swim-lane, describing the procedural algorithm of the transactional process showing hand-over interfaces lacking even Lean key metric or a bottleneck analysis. The hunt for Muda-reduction is not wrong and might be a valuable approach at the beginning, sensitizing employees to reduce non-value add work (NVA) and to concentrate on the value-add tasks (VA). In addition, the most applied and divulged concepts have been, apart searching for the seven wastes, implementing 5S workplace organization and Kaizen workshops. These statements might be considered heretic for some readers but they hit the essential reason why you hold this book in your hands. Indeed, the question arises how to implement SMED (Single Minute Exchange of Die) set-up reduction techniques or TPM (Total Productive Maintenance), cardinal tools in presence of production machines, but for sure not in the office environment dominated by human beings. However, also how to implement flow on pull in the office is not trivial, and therefore not usual. Indeed, the Lean transformation, such as Womack from the LEI proposed it in his five-step approach [3]:

- define value for customer
- map the process flow and eliminate waste
- introduce flow on customer pull
- empower people
- strive for perfection,

the third step "introduce flow on customer pull" has never, or let us say very seldom, been implemented in transactional environment. Further, the proliferation of several dedicated Lean approaches for banking, government, etc. [4]—let us call them Lean X—rarely have shown a break-through application and success, in my honest opinion. However, if transactional processes in the office are considered to be the same as transformation processes on the shopfloor, if ever possible, the same TPS optimization concepts and general production laws apply. In recent time, the cost pressure has seen some banks going to implement basic elements of standard works or even office space rearrangement based on industrial work cells. This shows that the cost pressure is high and the consciousness to go beyond merely explicit and obvious Muda-reduction has increased in the service industry. Such as in industrial environment, this makes it necessary to implement industry-paradigmatic waste-free processes designed to required quality also in the office environment. Again, also Sergio Ermotti, CEO from Swiss bank UBS even stated: "we have to industrialize the banking system" [5]. Therefore, we propose here an integral conceptual approach to the office Lean transformation; a new way how to industrialize the office environment to reduce Muda and gain in competitiveness in addition to the imminent transfer of Industry 4.0 concepts to transactional processes.

Further, the critical customer requirements, often called voice of the customer (VOC), for the services are the same as in the industrial environment. They can be reduced to speed, punctuality (we will call it on-time-delivery), and quality. To keep a business viable and sustainable also the economic return, commonly called profit, has to be integral part of the business equation. These four basic requirements form the SPQR model [6, 7]. These basic SPQR requirements can be considered to be time-invariant and constitute a sort of a minimal axiomatic multi-objective system,

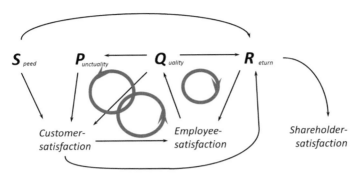

Fig. 1.1 SPQR model showing systemic effects between the cardinal objectives (from [7])

which has to be observed in any case, also in service industries, to be successful in business (Fig. 1.1).

This book tries to bridge the often-applied too simplistic approach of Lean Office books by searching for Muda according to the seven wastes. Indeed, when LSS OPEX initiatives start in transactional environment, independent of the selected approach, the seven wastes (Muda) are at the center of attention. Lean is often explained and interpreted with reducing waste in form of non-value add tasks. In Fig. 1.2, the seven original wastes are specifically detailed for the office environment. From all these Muda, waiting might be the biggest waste; to reduce this type of waste a different work organization has to be envisaged. Waiting is caused by different other type of Muda and caused by incomplete information as well as searching for additional input, too many open files, but also by motion due to retrieving documents. Therefore, Gemba walking the offices has been applied primarily; such as on shopfloor also on office-floor a lot of Muda can be detected in this way and can be eliminated. Representing graphically the process by swim-lane, showing the multiple handovers, additional improvement potential might be discovered. By applying 5S workplace organization technique, additional improvement can be obtained.

However, often Lean Office initiatives end with 5S introduction. If for such type of OPEX initiative a huge organization has been put in place, the cost may have exceeded the benefits and the initiative is considered to have failed. Not the initiative has failed but the management approach! In such a case, Lean has not been exploited and the major improvement potential has not been discovered.

To exploit Lean, we have to think out of the office box; we have to industrialize the office environment and to adopt shopfloor similar techniques to boost productivity. Process lead-time (PLT) and throughput, aka exit rate (ER), are central indicators to measure performance of a process, as we will see later. If ever possible, we have to strive to implement a single piece flow (SPF). Although it may sound silly or even impossible, we have to try. To try means to have understood the "real" TPS beyond explicit Muda elimination, but also, how to adapt to specific characteristic of the office world. The exploitable potential of Lean applications is shown in Fig. 1.3. Common waste-oriented Lean office approach (red line) is not able to exploit the improve-

Transpor-tation	Movement that does not add value	Retrieving or storing files	**O**ver-Production	Generating more information then the customer needs right now	More information than the customer needs
		Carrying documents to and from shared equipment			Working on next weeks items.
		Taking documents to another person or location			Creating reports no one reads
		Going to get signatures			Making extra copies
Inventory	More information, project, material on hand than the customer needs right now	Files waiting to be worked on	**O**ver-Processing	Efforts that create no value from the customers viewpoint	Creating reports
		Open projects			Repeated manual entry of data
		Office supplies			Use of outdated standard forms
		E-mails waiting to be read			Use of inappropriate software
		Unused records in the database			
Motion	Movement of people that does not add value	Searching for files	**D**efects	Work that contains errors, rework, mistakes or lacks something necessary	Data entry error, typo's
		Extra clicks or key strokes			Pricing error
		Clearing away files on the desk			Incomplete Forms
		Gathering information			Missed specifications
		Looking through manuals and catalogs			Rework Loops, Lost Documents
		Handling paperwork			
Waiting	Idle time created when material, information, people or equipment is not ready	Waiting for...	**U**nderutilized People	Untapped or misused human resources	Unclear job descriptions
		Faxes			Un-engaged professionals
		The system to come back up			Ambiguous goals & objectives
		Copy machine			Specialist not utilized in their area
		Customer response			
		A handed-off file to come back			

Fig. 1.2 The office Muda "TIMWOODU" [8]

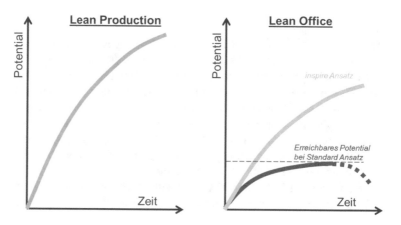

Fig. 1.3 Comparison of exploited Lean potential in industry versus office (from [8])

ment potential of industrial Lean transformation (blue line). The Inspire-approach (green line) tries to unchain the full potential of Lean also in the transactional office environment.

The aim of this book is to give a consistent and systematic approach to a revolutionary and sustainable office transformation. Several books have already been published in the domain of Lean Office, e.g. [9–13], list absolutely not pretending to be exhaustive and without any judgment. They hardly can be comparable with the description of the original TPS [2] or its interpretation, e.g. [14]. However, some shy appearance on the internet of office cell examples show that the issue has emerged. Further, the lack of a rigorous scientific approach of the Lean literature has even leading the author to formulate ex-post a manufacturing-based theory interpretation of Lean [7]. Indeed, Lean is much more than reducing waste, Lean is an alternative production theory to the existing Western models. However, the scarce success of Lean applied in the office compared to the industrial application shows that something does not work (see Chap. 2). Nevertheless, the office world represents today nearly 80% of employment, i.e. the major cost of transformation [15]; this improvement potential cannot be neglected. Therefore, this book presents a systematic approach of a paradigmatic application of the industry-based TPS transposed to the office. It tries to industrialize the office environment, where many people are working in an often not standardized manner at its own discretion—a huge improvement potential. In addition, we will not apply the usual consultant-based "tool-picking-approach", which may derive from the two-pillar temple house representation of the TPS. We will base on a new mono-pillar representation of the TPS [7], which stresses the systemic character of the tools interaction according to Fig. 1.4. Whereas this approach is mandatory in industrial cell design, it is highly recommended also in office process design.

In the following, at first we will enter the topic why so many OPEX (Operational Excellence) initiatives have failed in the office environment. The content of Chap. 2 corresponds therefore to the inciting incident of a play in classical literature. Further, Chap. 2 enters into the secret of the TPS. It is essential to understand the main concepts behind this most performant production theory in its original industrial environment to be able to transpose the full potential going beyond the concept of the seven wastes. It tackles also the intrinsic difference between shopfloor and office, basic reason regarding the difficulty to transpose the whole potential of the TPS. Further, it models the office system, i.e. how transactions and tasks are performed in an office. The complexity, or better the multitude of different and often not structured situations, need to enlarge the approaches contingent to the situations. The chapter is enriched by presenting the Relational Office Model, which helps to reinterpret the TPS adapted to the office world. Chap. 3 will deal with the basic TPS-tools and how to adapt their logics to the rational of a transactional office environment; this always keeping in mind the final aim to industrialize the office world, in order to increase productivity. It defines and structures the activities in the office obtaining a taxonomy of definitions, which helps to introduce systematics and rigor in the office world. It tries to formulate manufacturing-based principles transposed to the office environment. This helps to structure the office work and to establish more rigor as well as reproducibility to establish and manage the office tasks and execution of transactional processes. Based on the tools presented in Chap. 3, the Chap. 4 reveals how to implement a lean office cell based on cellular manufacturing concepts of industrial application. The implementation of an office cell is at the

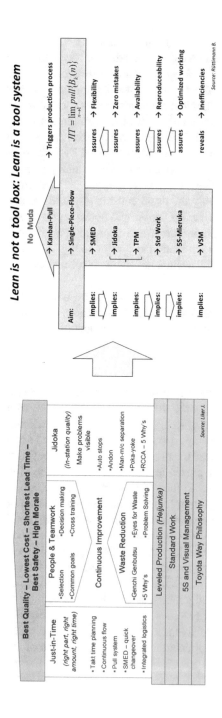

Fig. 1.4 From the Toyota house to a systemic Lean model (simplified didactic representation)

center of improved and Muda-reduced working boosting transactional productivity. Chapter 5 will explain how to introduce Lean in the office; i.e. how to manage daily a lean office, how to apply continuous improvement, and how to deploy and roll-out Lean Office across the whole company. And finally, Chap. 6 will address some critical thoughts and give an outlook to the forthcoming new digitalization era.

References and Selected Readings

1. Rüttimann, B.G.: Discourse about linear programming and lean manufacturing: two different approaches with a similar, converging rational. J. Serv. Sci. Manag. **8**, 85–91 (2015)
2. Ohno, T.: Toyota Production System—Beyond Large Scale Production. Productivity Press, New York (1988)
3. Womack, J.: Lean Thinking. Free Press, New York (2003)
4. Netland, T., Powell, D.: Routledge Companion to Lean Management. Routledge, New York (2017)
5. Inside Paradeplatz: UBS-Industrialisierung: Tod oder lebendig? News of 7 Dec 2014. inside-it.ch
6. Rüttimann, B.G.: The Central Importance of Quality, ALUMINIUM 77, 7/8. Giesel Verlag, Hannover (2001)
7. Rüttimann, B.G.: Lean Compendium—Introduction to Modern Manufacturing Theory. Springer (2017)
8. Inspire Academy: Lean Office Training, The Industry-Transposed Office Revolution—A Step Beyond Learning to See, Inspire AG (2018)
9. Kamiske (Hsgr.): Lean Office—Grundlagen, Methoden und Umsetzungsleitfaden (Pocket-guide), Hanser, München (2019)
10. Martin, J.W.: Lean Six Sigma for the Office. CRC Taylor & Francis, Boca Raton, FL (2009)
11. Locher, D.A.: Lean Office and Service Simplified. CRC Taylor & Francis, New York (2011)
12. George, M.L.: Lean Six Sigma for Services. McGraw-Hill (2003)
13. Locher, K.: The Complete Lean Enterprise—VSM for Office and Services, 2nd edn. CRC Taylor & Francis, Boca Raton, FL (2016)
14. Liker, J.K.: The Toyota Way, 14 Management Principles from the World's Greatest Manufacturer. McGraw-Hill (2004)
15. Fischer, U.P., Rüttimann, B.G., Stöckli, M.T.: Aufstieg und Fall von Six Sigma oder warum heute Lean wichtiger ist, iO new Management, Axel Springer Zürich (2011)

Chapter 2
Insights and Prerequisites for Lean Office

It might appear blasphemic to begin a book with a negative statement; however, we have to remain realistic: to introduce Lean in an office environment is much more difficult than introducing Lean in non-automotive manufacturing industries. This for two reasons: firstly, the work content of transactional processes is much less deterministic in most office applications and routines than it is in manufacturing processes, and secondly, due to the stochastic characteristic of office transactions, the execution is mostly left to the discretion of the employee. As a consequence, it is difficult to set common standards or even prescribed time slots to certain office activities such as it is the case in industry. Further, having white-collar employees generally a higher scholastic education compared to blue-collar employees, they often show more resistance to prescribing them the way how they should work; they like the non-strictly prescribed way to work leaving them more freedom—and, as a consequence, potential possibility for Muda. However, this is a difficulty, which can be mastered by understanding the intrinsic nature of an office environment; the real issue is managing the Lean transformation of an office—indeed, this is a change management issue. What can go wrong? This topic has been analyzed e.g. in [1], from which we will take some excerpts.

2.1 Reasons for LSS OPEX Initiatives Failures

Generally, also in industrial environment OPEX (operational excellence) initiatives have been failing. What are the reasons? The reasons often are linked to two facts [1]:

– Management incompetence due to ignorance about Lean Six Sigma (LSS) topic
– Choosing the wrong OPEX approach apt to the contingent situation.

We will not enter in the first topic, which can be solved by training management executives to become knowledgeable; we will focus on the second topic. Indeed, OPEX is not equal to OPEX—different approaches exist. Figure 2.1 shows how the today predominant LSS DMAIC problem solving approach proposed by the George

© Springer Nature Switzerland AG 2019
B. G. Rüttimann, *Transactional Lean: Preparing for the Digitalization Era*,
https://doi.org/10.1007/978-3-030-22860-6_2

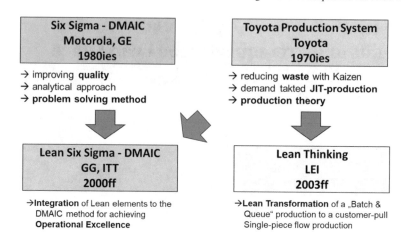

Fig. 2.1 Origin of Lean six sigma DMAIC and Lean thinking

Group [2] originates from the original Six Sigma DMAIC (Define, Measure, Analyze, Improve, Control) method integrating Lean elements of the Toyota Production System (TPS) [3] as well as [4]. Lean on the other hand is a recipe how to implement the TPS in Western manufacturing plants [4]. Indeed, Toyota developed and perfected their production system during 30 years; we do not need 30 years to do the same. The LEI approach bases mainly on the "present state—future state—action plan" transformation approach, transforming a B&Q-push (batch and queue) manufacturing system into a SPF-pull (Single Piece Flow) production system, transformation approach developed and proposed by the Lean Enterprise Institute [4]. The pure Six Sigma approach is treated e.g. in [5, 6] and the TPS is described e.g. in [3, 7].

Whereas this "present state—future state—action plan" Lean transformation approach is widely applied in the USA, in Europe rather the Carthesian "Lean introduction" approach is used which bases on the Toyota house. This approach is mainly applied by European consultants. They begin scholastic-like with the introduction of 5S and shopfloor management, introducing the Kaizen philosophy, which we have been calling continuous improvement, searching for Muda by Gemba walking. Recently this approach has been supported more and more by VSM (Value Stream Mapping). Nevertheless, the DMAIC approach and the Lean approach have been experiencing a different evolution (Fig. 2.2). Indeed, after a fulminant expansion of LSS DMAIC in Europe (especially Switzerland), the pure Lean approach divulged rapidly [9]. A study carried out in 2012 revealed, that it had a different diffusion of the approaches among manufacturing and service industries in Switzerland [10]. Today both approaches are applied in parallel.

Although all approaches are OPEX compatible, or better, suited to reach operational excellence, there are different pros and cons to be aware. Whereas the DMAIC approach is always applicable, it has its forces at the beginning of an OPEX initiative, especially if the company has serious financial problems, often called "burning

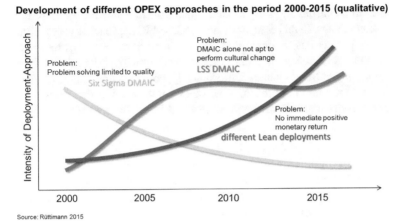

Fig. 2.2 Temporal evolution of different OPEX approaches in Switzerland [8]

Deployment	Advantage	Disadvantage	Requirements
LSS DMAIC	• General problem solving method • Structured training • Immediate dissemination of Lean Six Sigma culture • Fast financial return	• Gate reviews sometimes cumbersomely • Exhausting improvement potential • No sustainable change of culture	• Coaching of first project teams by experienced consultant is recommended • Suitable project selection is neccessary
Lean Introduction	• Sustainable change of culture starting from shopfloor • Kaizen-team based • Beginning with 5S	• Slow setup • High cost and initially lack of financial benefit • Quality not priority	• Requires patience • Needs long consultant assistance if knowledge is not internalised fast
Lean Transformation	• Fast conversion of production to Lean (from Push to Pull) • Immediate visible success	• High risk of initial production failures • Big cultural shock • Quality issues not fixed	• Requires strong leadership of management • Has to be accompanied by consultants

Fig. 2.3 Comparative analysis of different OPEX approaches [1]

platform". In fact, every DMAIC project gives an immediate quantifiable result and therefore economic return, whereas a classic "European Lean introduction approach" beginning with 5S and shopfloor management costs a lot of consultant fees with a limited financial return. On the other hand, however, DMAIC will not be able to change the culture of the company; for that, the pure Lean approach is necessary. Indeed, DMAIC is a problem solving approach whereas Lean is a way to run a company. The advantages of the three main approaches and the traps to be vigilant are summarized in Fig. 2.3.

Figure 2.3 shows, that there is not one universally applicable approach for OPEX, the approach should be selected in accordance to the specific contingent situation in which the company stays. A company not suffering negative financial results may apply the "Lean introduction" approach, which one is for sure not apt if the company suffers heavy losses. In this case, only the LSS DMAIC approach is very recommendable. Every project has its quantified benefit. In addition, it brings basic elements of Lean, but also Six Sigma, techniques to every employee laying the foundation to the following cultural change. Unfortunately, consultant companies usually have their single preferential approach, which they try to sell. For that reason, management executives should be knowledgeable about LSS to select the right consultancy company.

The question now arises: does an ideal OPEX approach exist?

To overcome the pitfalls identified in the previous section it is recommended to apply a holistic OPEX approach. If the company presents a bad financial shape it is highly recommended to apply first the LSS DMAIC approach, and switching then to a systemic Lean transformation of the production system and introducing in parallel the continuous improvement culture of Kaizen. Many companies made the mistake to stay too long with DMAIC. However, if a company is in bad financial shape one cannot start with the shopfloor continuous improvement approach, taking too long time to achieve the benefits of the cultural change.

In Fig. 2.4 is shown the different phases of the OPEX journey following a sort of multi generation plan (MGP) for the OPEX transformation proposed by Inspire. If a company is sound, it can start directly with the Lean transformation and lean shopfloor management. However, the DMAIC is a general problem solving approach, which can also be applied when the Lean Transformation has been completed.

By the way, also medical doctors are applying a DMAIC approach. In fact, when a patient comes to the MD he first asks the patient what is the pain (Define phase means problem description). Then he visits the patient compiling the anamnesis and ordering some blood test (Measure phase means establishing the base line). Based on the findings he formulates the diagnosis (Analyze phase means to find the root causes). After having identified the root causes of the disease, the MD prescribes a therapy (Improve phase means to find the optimal solution). At the end, he invites the patient to return in some weeks to see the results and to adjust eventually the therapy (Control phase means to assure a sustainable process). This MD consultancy process is a remedy; however, if the patient gets rid of the sickness depends how he is conducting his way of life. The same is also valid for a company. DMAIC can fix problems but to have sustainable results the management approach has to be changed, such as following e.g. the Lean philosophy management approach. This shows that there is not a standard approach for OPEX; the correct OPEX approach has to be chosen from a comprehensive tool/approach system to get the best result in presence of the contingent situation of each single company.

There is another trap linked to LSS introduction into the office environment: this is less linked to selecting the wrong deployment approach but rather linked to not exploiting the full potential which Lean and Six Sigma entails. Although Six Sigma has a strong tie to manufacturing industries and less to transactional service industries, the concept of quality is omnipresent, also in the offices. Quality is therefore also a main topic of office processes and it has compulsory character. Other than usually

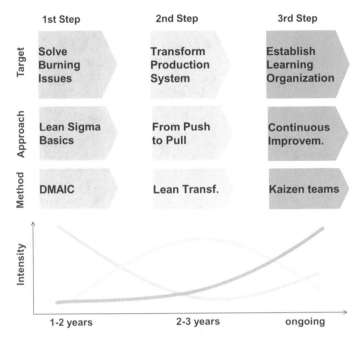

Fig. 2.4 The Inspire® approach for industrial OPEX transformation (*). (*) Inspire AG is a technology transfer institute of the ETH Zurich, the Swiss Federal Institute of Technology

believed, Six Sigma has its fix and important place in transactions; however, instead of talking of quality we will talk of conformity, that sounds more appropriate for administrative processes.

Nevertheless, we are conscious that Six Sigma, i.e. quality, is only one topic in production theory, whereas Lean is itself a production theory comprising the aspect of quality called Jidoka. From that point of view, Lean clearly dominates the Six Sigma approach. As we have seen, the LEI Lean transformation approach is structured in

– Define the value for the customer
– Select the value stream and eliminate Muda
– Implement flow on customer pull
– Empower employees
– Strive for perfection.

Whereas Lean office teams have no problems to search and eliminate Muda explicitly, having also made big progress in increasing office productivity, generally, Lean office initiatives slow down when coming to implement a flow on pull value stream principle. This is not surprising, indeed, how to implement practically in an office world a pull principle based on supermarkets? In addition, the problems increase when to interpret and implement TPM (Total Productive Maintenance), or SMED (Single Minute Exchange of Die), or a Heijunka-box leveling of activities. How to

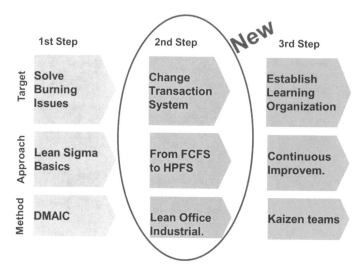

Fig. 2.5 The Inspire® approach for Lean Office OPEX transformation

interpret and implement these tools in an office environment? It is therefore not surprising, that most Lean office initiatives seldom have been going beyond eliminating explicitly Muda without discovering the benefit of additional Lean advantages. And exactly this is the topic that we will tackle in this book: the industrialization of the office environment in order to boost productivity.

In Fig. 2.5, we see the Inspire approach for the office lean transformation, derived directly from the industrial approach of Fig. 2.4. Also for Lean Office a DMAIC problem solving approach is recommended at the beginning. Indeed, LSS DMAIC has two benefits: first, it helps to fix immediate existing issues by improving quality and flow of operations; second, it helps to divulge the first notions of Lean techniques and the sense of urgency to reduce non-value add (NVA) waste in the processes. Fixing immediately burning issues is important; no employee understands the rollout of a huge transformation project if daily recurring problems are not yet fixed. However, the second step of the MGP is different from industry. Why it is different, we will see in Sect. 2.3. Solving the Gordon knot will help to industrialize the transactional office processes gaining productivity.

2.2 Understanding the Toyota Production System[1]

To industrialize a transaction-based office environment becoming lean we have first to understand industrial lean manufacturing systems. Understanding the Toyota production system (TPS) means to have understood high performance production

---

[1]Part of this section is derived from the paper [11].

systems. The TPS differs fundamentally from classic Western batch and queue (B&Q) push manufacturing systems with long process lead times (PLT). The traditional office work can to some extend be associated to a B&Q manufacturing mode building WIP along the process. The Lean Office deployment failures can also be attributed to some extend to non-having understood in depth the TPS. Indeed, the TPS is often identified with Muda reduction, and that is by far too simple, although not wrong. In this section, we will not only scratch at the surface of the production-theory aspect of the TPS, but we will go beyond Muda Gemba walking. For further details regarding TPS consult e.g. [3, 7] and for a more academic treatment consult [12].

If you ask employees, but also consultants, what does Lean mean, there is a high probability to get the answer: a Muda-free process. Muda is defined as non-value add time for which customers are not willing to pay. Muda is the Japanese word for waste, where Toyota defined the seven wastes as

– Transportation
– Inventory
– Movement
– Waiting
– Overproduction
– Over-processing
– Defects.

In addition, in the Western world also unused skills of employees is considered to be the eights waste, because usually, to the contrary to Japan, the employee is considered to be a cost factor which has only to obey and execute work-content according to standard operating procedures (SOP). Indeed, Toyota has developed the employee-based Kaizen (continuous improvement) approach to improve the manufacturing system. It is the employee, who executes every day his work, who knows best how to organize and to make process steps more efficient by eliminating Muda. Toyota really empowered blue-collar workers to help to develop and perfection the TPS.

Another characteristic of the TPS is Mieruka (free translated as visual management). Visualizing and sharing necessary information, e.g. the status of the production lines by Andon lights or the achievement of production schedulings by boards, is central in the TPS to align information spreading to employees. Shopfloor boards gather key information to be shared among employees at the shift handover. These shopfloor information boards, also called master boards or Kaizen boards if used to apply Deming's PDCA cycle, are used to discuss opportunities for improvement, tracking ongoing improvement projects, or to solve problems with the help of the Ishikawa diagram ("fishbone") and applying the 5-Why technique. Going to the Gemba or Genchi Gembutsu means "go and see, go and study, go and solve", i.e. one has to go oneself to the shopfloor where the transformation happens in order to get a non-distorted impression of the situation.

This technique has often been applied in transactional industries, especially banks and insurance companies, searching explicitly for Muda, supported by value stream mapping (VSM). "Learning to see" has become a central element in the Lean journey [13]. However, the Lean Office transformation in service companies has seldom gone beyond swim-lane diagrams, explicit Muda hunting and elimination, and the introduction of 5S workplace organization. In fact, how to implement SMED or TPM?

Another limiting factor has been the TPS house itself (temple model) with which Toyota explained to suppliers and customers their manufacturing system. Toyota discovered already in the eighties that to improve further the TPS it had do involve suppliers and customers, integrating in-bound and out-bound logistics into their system. Toyota realized already at its time what is one of the purpose of today's Industry 4.0: interconnecting the supply chain with internet of things (IOT). To talk only of the TPS is therefore reductive; in fact, one should talk of the Japanese Production System. In Fig. 2.6 we see the original version of the TPS model which looks fairly different than today's reinterpretation. On the internet exist dozens of reinterpretations of the classic TPS model, simplified or distorted, but all have the same deficiency: it is a descriptive model. Many tools are mentioned in the TPS model and if one beliefs to implement Lean just by picking some single tools and use it at its own convenience, one is wrong. Okay, that is not completely wrong but limiting. Indeed, as already shown in Fig. 1.4 of Chap. 1, an alternative systemic mono-pillar representation suits better for didactic purpose to show the holistic functioning of the main TPS tools than the original representation shown in Fig. 2.6 [11]. All tools have to be implemented to assure the perfect functioning of a single piece flow (SPF). All these tools are necessary to implement a perfect JIT production system; if one is left, this could compromise the perfect working of the system. Indeed, the TPS cannot only be interpreted as Muda-free management approach, the TPS is the most sophisticated production theory. The theory content of the TPS will be explained on the next pages entering into the manufacturing-theory aspects of Lean.

As the acronym TPS reveals, the TPS is a system. Generally, a system is composed of different elements which alone do not make sense but together convey to a common aim. In our case, the TPS tools are necessary to assure a demand-pull JIT SPF to guarantee on time delivery (OTD). This comprehensive system is synthesized in Fig. 2.7 showing the functioning by increased production complexity.

Figure 2.7 shows the concept of the TPS working mechanism and reflects the reality of a complex manufacturing plant. For introductory purpose an easier didactic representation of the TPS is shown in Fig. 2.8, which reveals the basic working mechanism of the TPS, i.e. the required main techniques, often called tools, to implement a flawless SPF of a transfer line or a manufacturing cell. Figure 2.8 shows that all the main Lean tools have to be put in place to implement a SPF.

A real manufacturing environment, however, is made of several products needing several machining operations performed in different cells. These cells C_j or better shopfloor ateliers comprise usual processing-technologies such as sawing, machining, grinding, welding, heat treatment (often batch operated), surface treatment, assembly and painting. The simplest production case is the *mono* product manufacturing, ideal for the introduction of a SPF to reduce PLT. This is done by minimizing WIP with a paced production line meeting the required TR. To guarantee the correct pace of the line, techniques such as 5S, standard work, TPM, Jidoka, balancing have to be put in place. When *multiple* products are manufactured within the same cell (mixed product cell manufacturing), still maintaining a SPF,

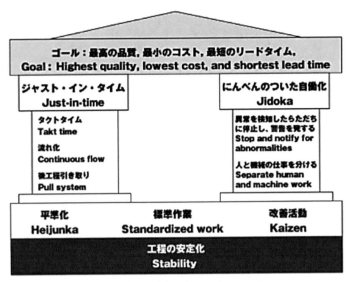

Fig. 2.6 The original TPS model from mid-1980s (*Source* internet: Kaizen Express, John Shook and Toshiko Narusawa)

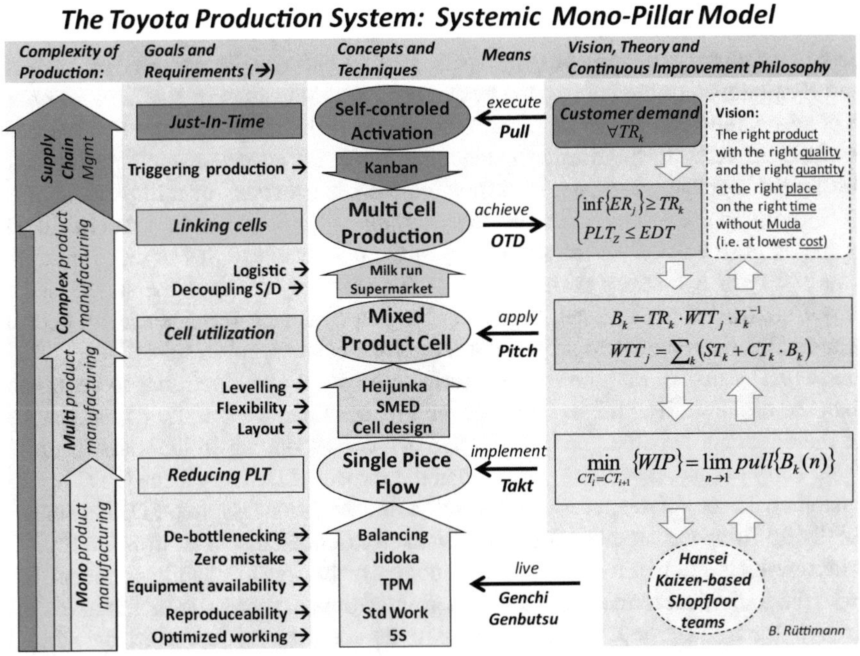

The Toyota Production System: Systemic Mono-Pillar Model

Fig. 2.7 The systemic production-theory aspect of Lean beyond Muda and Kaizen (from [11])

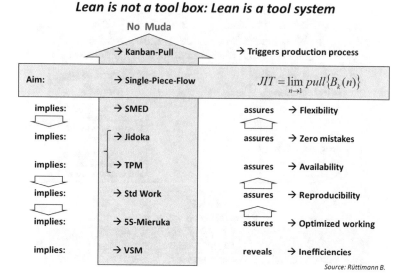

Fig. 2.8 Didactic mono-pillar model of the TPS

a further complication has to be mastered. Indeed, the batches B_k have to be sized to the Takt Rate TR_k of the product k and the Workstation Turnover Time WTT_j or even Cell Turnover Time CTT of the cell C_j if a JIT delivery of several products is required. The applied techniques for this purpose are cell design, SMED, and Heijunka box scheduling. However, the production situation is often a complex product-manufacturing environment comprising different processing-technologies in different cells. In this further extended complexity, several manufacturing cells are linked together via strategic buffers, called supermarkets or milk-run supplied POU (point of use) buffers. Such buffers decouple the non-synchronized demand (D) of the downstream cell to the supply (S) of the upstream cell due to different cycle times (CT) of operations within the cells. The replenishment of the supermarkets is self-controlled via Kanban, triggering the production before a stock-out occurs. And finally, the requirements to be observed for a customer on-time-delivery (OTD) are two. The first is, that the smallest Exit Rate ER_j of all cells C_j has to be greater than the required TR, and second, that the Process Lead Time PLT_Z of the last, i.e. of the customer "visible" processing step Z—corresponding to the manufacturing order entry point—have to be shorter than the Expected Delivery Time (EDT) of the customer. These are the necessary and sufficient conditions for an OTD. This means finally fulfilling a customer JIT supply. Such an extended model is shown in Fig. 2.7 which reflects the mathematical full induction or backward-chaining concept (i.e. from the individual to the general view), going from the mono-product manufacturing, via the multi-product manufacturing to the complex product manufacturing. It represents the increased complexity of shopfloor logistics.

All these interactions are shown in the cognitive model of Fig. 2.7, a comprehensive, systemic view of the modular construction of the TPS-Lean Model, showing also the rationale for each logical manufacturing complexity. The related theory is described in [12]. It clearly states the goal of the concepts and which technique needs to be applied in order to satisfy the requirements to achieve the overall goal. In addition, the shopfloor continuous improvement philosophy is shown too (Kaizen teams), which represents the daily small improvements on all stages. Indeed, the final vision of "the right product with the right quality and the right quantity at the right place on the right time without Muda" needs the implementation of all TPS-techniques, which transform the underlying theory into action, will implicitly lead to eliminate Muda. Western companies probably have the impulse to add "at the lowest cost" to this final vision, what, however, is not necessary, since achieving this vision implicitly leads to lowest cost.

For our purpose to implement the Lean Office we will not base on the exhaustive model representing the systemic-based Toyota manufacturing system shown in Fig. 2.7 but we will use the simplified didactic model shown in Fig. 2.8.

2.3 Understanding the Office World[2]

In Sect. 2.1 we have seen some reasons why Lean Office initiatives failed. In Sect. 2.2 we have seen in what the TPS consists of and what Lean really is. It is much more than merely reducing Muda; it is a perfect production theory which, when correctly applied, will implicitly eliminate Muda. To transfer this theoretic framework to the transactional process environment we have at first to understand what the differences between industry and services are. In this section, we will analyze the intrinsic differences between shopfloor and office.

Why is it so difficult to deploy Lean in an office environment? This is linked to the characteristics of the specific manufacturing type, i.e. the different operations characteristics of transaction services compared to manufacturing industries. Indeed, the transaction object, i.e. the phenotypic product, as well as the process execution, i.e. the way how the product is manufactured or the service is performed, but also the work governance, i.e. how the transformation is controlled, are intrinsically very different between industry and services. Figure 2.9 gives an overview of the differences of product characteristics in manufacturing and service industries.

Interesting is the difference of product definition, variation, and quantity. In industry the product usually is a defined physical object with certain characteristics; the product usually is an off shelf product, eventually with possible customization. The product is manufactured identically again and again in an appropriate batch-size. We can therefore talk of an "identical repeatability", although it seems to be pleonastic but it is not in this context, as we will see. In transaction industry, however, the product is a template or a one off task. We leave apart the latter case. The template is a

---

[2]This section is largely based on the paper [14]

	Shopfloor	Office
Transaction object	Physical product	Service, file, information
Morphology	Tangible, visible	Intangible, invisible
Product definition	Determined according to defined specification	Ranging from parametric defined template to VOC
Product variation	None	Virtually unlimited
Product quantity	Virtually unlimited batch size	Single transaction batch size
Product non-conformity	Visible with defined out of control procedure	Break, often invisible, undefined mitigation
Degree of control	High (directly influenced)	Low (externally influenced)
Predominant actions	Executing	Analyzing/planning/executing
Make to	Make to stock possible	Make to order
Assets	Mainly product specific equipment, capital intensive	No task specific equipment, IT is integrating backbone
Set-up cost	May be relevant	Low, but task switch-over

Fig. 2.9 Differences of transaction object

"parametric product", a template that has to be compiled for every customer order. Every order has a different address field, a different transaction amount in the case of a credit, a different duration, different financial conditions, etc., i.e. each "product" is different and it is performed only once. It is a transaction with batch-size one. We can therefore talk of a "formal repeatability", meaning repeatability of the process but not of the content. This difference will lead us to talk not anymore of a single piece flow (SPF) but we will talk on the office-floor of a single file execution (SFE).

In Fig. 2.10 are shown the differences in process execution between shopfloor and office-floor. Due to the different task description, indeed, in the office less the task than the desired output is described, also the task characteristic is less structured and therefore also the transformation model is rather relational. Also the value-add content is different, comprising in the office also thinking and interacting if customer-focus related, whereas on the shopfloor it is merely executing according to SOP, simply said.

In Fig. 2.11 are outlined the differences of work governance between manufacturing and office. The main difference is given by the consciousness of thinking logic. In manufacturing, due to the high repetitiveness of tayloristic-fragmented work in high performance lines, the operations are executed by the operator in a subconscious way what is represented by the concept of Kata; this corresponds to a wired logic. In the office, due to the change of parametric content of the template, the tasks are executed in an explicit conscious way; this corresponds to a programmable logic. This has a direct consequence on the control of the execution as well as of the supervision of the work progress, which is much easier on the shopfloor needing less flexibility.

	Shopfloor	Office
Principle orientation	Process-centric (task-oriented)	Output-centric (result-oriented)
Transformation model	Procedural type	Relational type predominant
Task description	Detailed SOP including time to comply with customer takt	Approximate, rather description of output to be performed
Task characteristic	Very structured, no degree of freedom, no alternatives	Less-structured, allows execution discretionality, many alternatives
Task content (width)	Simple, confined (narrow, to be takted)	Complex, comprehensive (large content, difficult to be takted)
Waste, Value stream	Visible	Mainly invisible
Value add	Usually transformation time	Includes think, write, listen, talk
Process flow	Ideally 1-piece flow	Mainly push on „boss"-pull
Process concept	Usually sequential steps	Random access to ressources
Process aim	Balanced and takted flow	Maximize parallelization

Fig. 2.10   Differences in process execution

	Shopfloor	Office
Kahneman thinking logic	System 1 (subconscious)	System 2 (explicit)
Flexibility logic	Wired	Programmable
Execution liberty	Very low (SOP-driven)	High (or template-driven)
Person's function	Support to process	Main actor of process
Execution	Very structured	Contingent approach
Measurability	Easy	More difficult
Disturbance noise	Limited, mainly endogen	Heavy, mainly exogen
Control	Implicit (poka yoke, kata)	Difficult, because often hidden and high variety
Consequence	No degree of freedom assuring repeatable and reproducable quality	High degree of freedom how to perform the task to get the output
Out-of-control cases	Very limited	More frequent
Focus	Efficiency	Effectiveness

Fig. 2.11   Differences in work governance

These differences have to be taken into account when transposing the Lean tool system to the office conceiving a Lean Office cell. Indeed, the Lean tools suit a deterministic environment. The stochastic variability of executing a template has to be taken into consideration, so where typical cycle times in the shopfloor are seconds or minutes, in the office they are minutes or hours. This time may comprise also interaction time with the customer to clarify an unexpected changed situation.

Let us now have a look at process types and related approaches.

Before we enter into the transfer of the Lean toolset to the office activities, we need to understand and characterize the desk job and to generate a model of it. The office process landscape can be grouped into the following three categories:

– Category Operational Processes (i.e. production and service generation)
– Category Support Processes (i.e. sourcing, HR, accounting, maintenance)
– Category Management Processes (as corporate governance, strategy development, risk management, budgeting).

The traditional Lean Six Sigma Community distinguishes, however, only between two types of processes:

– Manufacturing processes and
– Transactional processes.

However, this distinction is too simplified, as we will outline later. The transactional environment requires a further differentiation of all transactional processes. Indeed, solely the operational transaction processes are similar to the manufacturing processes and mainly for those we could apply continuous flow production considerations and realize cell design, but even this requires a re-interpretation of the original Lean concepts.

As transactional processes, we usually understand not only the operational, but also the service generation processes, and all office, administration, and supportive processes. The transactional process types are prevailing in an enterprise, and therefore, they require a particular attention. In general, the transactional processes are represented generally by swim-lane type graphics in order to visualize the interfaces, which represent hand-overs; for shopfloor production processes, however, typically value stream mapping charts are used, which focus on work in process (WIP).

In contrast to the production workplace the activities at an office workplace are manifold and much less repetitive. The differences can be summarized as follows:

– non structured versus structured (work of an employee that follows Q-handbook instructions)
– non controllable versus controllable (transparent insight for the line manager)
– unique versus repetitive (repeatability and learning effect of the task)
– manifold/complex versus monotonous/simple (work content)
– alone versus in team (work scope)
– complete versus part of a process (inter-functional labor division).

Indeed, the production activities are limited to a few, exactly structured and in terms of timing predetermined processes at one workplace in order to comply with the customer tact, such as

- operate
- monitor
- recognize
- feedback/act

which contain a very executing character, are supportive to the Jidoka-principle and need therefore a rigid "wired control logic" in order to produce quality. The activities in an office are more complex. Therefore, they need to be prescribed in a less deterministic way. Due to the large scope of the potential events, the work of the managers consists mainly in the following activities:

- confront, recognize
- think
- decide
- act/communicate (write, read, call, discuss)

and for the person in charge

- search documents/records (biggest non-value add position)
- reflect
- copy/scan
- fill-in
- conclude/execute
- inquire/check
- document
- archive
- communicate.

This supposes to a large extent a "programmable logic" in order to control the processes, although here, in some areas, a fixed wired logic would be required as well in order to achieve the intended quality. It is particularly important to reach and hold the required quality of processes and products, which is called reproducibility in manufacturing. Although office activities have a certain resemblance to the executing character of the production, the content and the provision of performance is clearly different. Whether an office activity contains more "thinking/deciding" or more "implementing/executing" content, depends mainly on the position of the employee. While the repetitive/executive activity dominates in production, we find a stronger verticalization of hierarchy and work content in the office area. The higher the position, the more the work content is unstructured and exogenously determined, the more executing the position, the more structured and repetitive the activities are. Therefore, we mainly find on the lowest hierarchical level task-centric process orientation of the executing person in charge (e.g. back office) similar to the production that strongly contrast with the result-oriented view of a manager, an assistant or a staff person. These differences have already been summarized in Fig. 2.11, which outlines the governance of the execution logic. Very interesting is the approach of Kahneman [15]: The repetitive, automatically executed tasks in production can be considered, "System 1 Processes", which are controlled by the unconscious. In the

Toyota philosophy, they correspond to "Kata". The activities in the office require a higher consciousness in order to complete the tasks; the person in charge has to work dedicated and in a focused manner. This conscious acting corresponds to "System 2 Processes". Here we can see the importance of the human resource, the person is required for "exception handling", and solving any problems that often appear. In the industrial language we would call this mastering "out of control"-situations, i.e. recognizing a wrong accounting entry, interpreting and correcting it.

Hence, we observe in transactional environments various process types, such as:

– operational processes (executing value generating routine tasks)
– supporting processes (executing other routine tasks)
– execution processes (executing tasks that are not routine)
– decision processes (comparing solutions and quantifying results)
– planning processes (preparing an organizational implementation)
– problem solving processes (analyze symptoms and find causes).

In fact, only pure operational processes (hereunder fall the value adding transformational processes) as well as supporting processes (rather non-value adding administrative tasks like book keeping) can be compared with the procedural processes and even these only, if they have enough repetitive content. A characteristic of the process character may be:

– unique (little structured)
– repetitive (structured with a given sequence)
– iterative (as a special case between unique and repetitive).

Apart from the repeating processes, also the iterative processes are very interesting, as those offer usually a high improvement potential. The administrative processes fall mainly in repetitive process characteristics, the problem solving processes into unique, if they are not solved using standardized problem solving procedures and if the problem description is not structured.

Decision processes fall as well in the unique process category, if they are not part of a larger process, i.e. budgeting, which often have iterative and recurring characteristics, where the decision can be reduced to a simple "fixed-wired logic". Iterative processes should not be mixed-up with repetitive processes, although in some cases a similar work step can be repeated many times. An iterative process can be repetitive, if it occurs regularly (budgeting). A process must be a sequence of repetitive activities in order to be comparable to a classical procedural production process. Here one can apply a traditional Lean Six Sigma approach. In summary, the category of improvement techniques is determined by the process characteristics. The following dimensions have to be considered:

– Process frequency (unique vs. repetitive)
– Number of process steps (one vs. many)
– Recursiveness of the process steps (unique vs. multi times)
– Clarity of the output definition (unclear vs. clear).

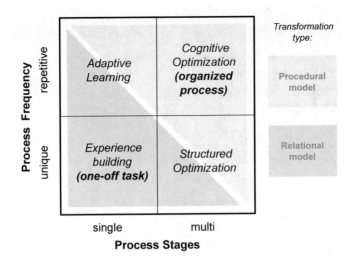

Fig. 2.12 Transformation type and improvement approach in function of process character

Figure 2.12 takes the first two dimensions and shows the spectrum of improvement approaches that sweeps from a "one-off task" (unique and integrated task) to an organized process (in the sense of a repetitive multi-step procedure). The applicability of the classical TPS and of the LSS approach for procedural processes is mainly in the upper-right quadrant; for all other, more relational processes, alternative approaches are required. The figure shows how variegated the process landscape in the office looks like.

However, as Prof. Martin Adam from FH Kufstein states, an alternative segmentation of processes should also take into consideration future changes due to increased digitalization. Such a segmentation might be

– classical routine processes, and
– knowledge-based processes.

Indeed, the first type of processes are subject to industrialization by robotic process automation (RPA) with softwarebots reducing human interaction. The second one might be heavily influenced by increasing artificial intelligence (AI) application.

Let us now have a look if it is possible to model the peculiarities of an office leading to formulate a specific office model.

As we have seen before, a comparison is necessary between production and transactional processes (operations, support, governance) in relation to the transaction object, the process characteristics and the work execution is necessary in order to understand the differences between manufacturing and office processes. The two processes differ strongly in product and information flow and particularly in the service provision. Although the support processes are implicitly or explicitly (Q handbook) defined as well, they show a much smaller degree of standardization of their activities

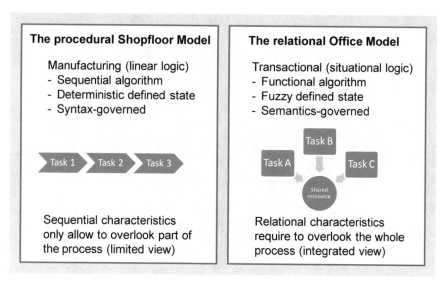

Fig. 2.13 Synoptic comparison—shopfloor versus office

in comparison to manufacturing processes and are much more subject to random-ness as a consequence of the freedom of execution and external influences. This is also related to the characteristics of the transactional object. The physical prod-uct is almost always defined based on customer specifications in a "deterministic manner", and has to be exactly repeatable ("identical repeatability"). In contrast, the performance of a service is determined "parametrically", and represents through the uniqueness of its specification at best a "formal repeatability", that relates to the repetition of the process, but not to the content. At best, the required information processing has a wide spectrum, which reaches from filling out a given form (i.e. structured IT-controlled order entry) to a completely unstructured "Voice of the Cus-tomer" (VOC) information (i.e. oral instruction by the line manager). Therefore, it corresponds more to a relational than to a procedural transaction. Another differ-ence exists in the capital intensity of the required infrastructure and therefore also the attention that is given to a maximal usage of the equipment by applying Total Productive Maintenance (TPM) and the OEE metric, KPI which is applied only to equipment and not to people.

The ROI key figure plays a minor role in service companies in comparison to industrial firms with a large asset base. In services, the productivity of the employees is key—not the productivity of the equipment as in the industry. Furthermore, due to the required decisions, any human activity represents a characteristic that is at the discretion of the executor and leads to the absence of "identical repeatability". Exactly this lack of "identical repeatability" in most of the processes will lead us towards a different modeling of the desk job, one that is more form-like (functional and relational instead of procedural), but also defined by the target instead of deter-ministically prescribing. Figure 2.13 highlights the relevant structural differences

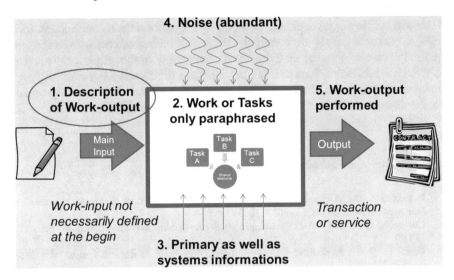

Fig. 2.14 The relational office model

between the production and transaction processes. It is well possible, however, that we also find some repetitive and structured processes in the transactional area that represent a typical procedural sequence comparable to those in the shop floor. In the focus of the reinterpreted office model stands a paradigm change from work-content oriented production activity to work-target oriented office activity. We could talk about reversed input-output modeling.

Because that we have to achieve in manufacturing a customer defined takt, work content of the processes is reduced to small repeatable elements that increase efficiency enormously (Taylor's division of labor). In the transactional processes of the office world, only a limited sequencing of the work, distributed among several persons, takes place. Even if a division is made, it has mostly functional character, the work content is not exactly defined and process alternatives exist. Therefore, even a "formal repeatability" may be missing. This means that the person in charge has a larger area of tasks and therefore some "enriched" job. This should also improve work quality, because less "hand-overs" happen that require information exchange. This is clearly necessary, as many repetitive work processes are identical, but the work objects, the information for every new order, are varying. The primary transaction object in the office is the file, the file is only a vehicle, which needs additional information as input for the production of the service itself or for performing secondary clarification processes. This discrepancy is the reason why the introduction of a takted pull system in the office would be inefficient, unnatural, difficult or even impossible to implement. That is why the TPS for service companies has to be newly interpreted, whereby Lean is getting a different meaning.

Figure 2.14 summarizes the intrinsic characteristic of the office world. One of the main difference consists how the task is described. Indeed, usually not the task is described but the output to obtain; how employees perform the transformation from input to output is not always clearly described (depending of course from the type of service industry) and often let to the discretionality of the employee. This way to work has functional character. In mathematics a function is a bijective mapping; however a relation is in mathematics not unique, which is therefore more appropriate for the description of an office; we call it therefore Relational Office Model.

References and Selected Readings

1. Rüttimann, B.G., Stöckli, M.T.: Reasons for operational excellence deployment failures and how to avoid them. JSSM, 612–618 (2015)
2. George, M.: Lean Six Sigma: Combining Six Sigma Quality with Lean Speed. McGraw-Hill (2002)
3. Ohno, T.: Toyota Production System—Beyond Large Scale Production. Productivity Press, New York (1988)
4. Womack, J.: Lean Thinking. Free Press, New York (2003)
5. Pyzdek, T.: The Six Sigma Handbook. McGraw-Hill, New York (2003)
6. Töpfer, A.: Six Sigma—Konzeption und Erfolgsbeispiele für praktizierte Null-Fehler-Qualität. Springer (2007)
7. Liker, J.K.: The Toyota Way, 14 Management Principles from the World's Greatest Manufacturer. McGraw-Hill (2004)
8. Rüttimann, B.G.: OPEX Deployment Ansätze im Vergleich: Gibt es den idealen Ansatz?, Presentation held at: Lean Six Sigma Dialog Summit, November 6, ETH Zürich, organized by Gesellschaft zur Förderung der Forschung und Ausbildung in Unternehmenswissenschaften an der ETH Zürich (2015)
9. Fischer, U.P., Rüttimann, B.G., Stöckli, M.T.: Aufstieg und Fall von Six Sigma oder warum heute Lean wichtiger ist. IO new Management, Axel Springer Zürich (2011)
10. Rüttimann, B.G., Walder, H., Adam, M., Stöckli, M.: Lean Six Sigma in der Schweiz—Explorative Studie zur Standortbestimmung. Inspire/ETH, SISE (2012)
11. Rüttimann, B.G., Stöckli, M.T.: Going beyond triviality: The Toyota Production System—Lean manufacturing beyond Muda and Kaizen. JSSM 9, 140–149 (2016)
12. Rüttimann, B.G.: Lean Compendium—Introduction to Modern Manufacturing Theory. Springer (2017)
13. Rother, S.: Learning to See. LEI/Cambridge Center (2003)
14. Rüttimann, B.G., Fischer, U.P., Stöckli, M.T.: Leveraging Lean in the office: Lean office needs a novel and differentiated approach. JSSM 7, 352–360 (2014)
15. Kahneman, D.: Thinking Fast and Slow. Penguin Books (2012)

Chapter 3
The Office-Adapted Lean Tool System

In Chap. 2 we have learned of what consists the TPS. We have seen that the TPS is not only a tool box but a tool system to implement a perfect working in-line quality generating single piece flow (SPF) based on demand pull, the final aim of the TPS. We have learned that the office environment differs from the shopfloor environment and that we have to reinterpret the TPS tool system. Based on the peculiarities of a transaction we will take the Lean tool system model of Fig. 2.8 and, where it makes sense, transform it into the conjugated Office Lean tool system (Fig. 3.1).

This is necessary, because the TPS tools are perfectly suitable for a discrete piece type of manufacturing and has to be adapted to the office transactions characteristics. If someone sais, that cannot work, it will for sure not work. Indeed, the concepts applied to perform a perfect transformation of inputs into output are always valid. Therefore, we can take the structure of the mono pillar TPS model but adapt the tools for the office transactions. In fact, also in the office world the aim is to perform a certain service as fast as possible; for that the one-file-execution is mandatory. However, in an office all employees are multifunctional and are required to perform different tasks. To do so, a fast job switch-over from one job to another job or routine is necessary; i.e. flexibility has to be assured. Further, the service provided has to comply 100% to customer requirements without rework by doing it the first time right; this corresponds to the quality aspect of the TPS. To transform inputs into output in an efficient way the team or the single employee has to be available and perform on a high level—always. Due to the fact that employees have to be interchangeable, it has to be assured that all employees are working in the same methodic way guaranteeing reproducibility of work. To work in optimal way without Muda the workplace has to be organized properly. For that visual management tools are necessary. The prioritization pull will assure that the right office process is triggered first in order to maximize efficacy. And what is with automation? The backbone "machine" in the office is the IT system. It represents the pivotal element in a transactions based environment, especially in view of big data. Finally, the key Lean tool, the VSM, helps to visualize the whole process and serves as central discussion mean to share the present state of efficiency and the improvement potential of waste reduction. In this chapter we will enter the main TPS tools and look how to adapt them for the

© Springer Nature Switzerland AG 2019
B. G. Rüttimann, *Transactional Lean: Preparing for the Digitalization Era*,
https://doi.org/10.1007/978-3-030-22860-6_3

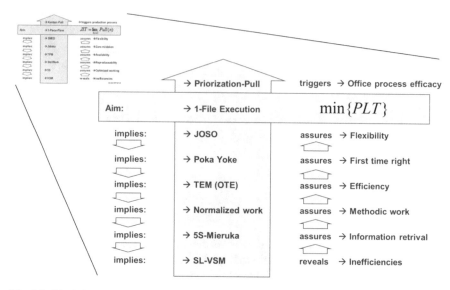

Fig. 3.1 The industry-derived and adapted Lean Office Tool System

office world. We will evaluate each single TPS tool and how to interpret it for the office environment to take benefit from the synergic interaction of the tool system. With that in mind we base our approach on the essential assumption, or better intention, to industrialize the office world [1] to gain competitive advantage by reducing non-value add Muda and increasing productivity.

3.1 Swim-Lane VSM and Lean Metrics

Although Value Stream Mapping (VSM) is not an original TPS tool—it has been developed by the Lean Enterprise Institute (LEI)—it is perhaps the most important Lean tool to describe and model simply industrial processes and reveals where waste is generated. Indeed, it maps the process sequence and the cycle times (CT) of each process step, the raw material (RMI) and finished goods inventories (FGI), as well as the work in process (WIP), and also the information flow. In addition, it presents a bottom line showing the process lead time (PLT) by classifying value-add time (VAT or simply VA) and non-value add (NVA) tasks, as well as the ratio of VAT and PLT which represents the process cycle efficiency (PCE) aka Lean indicator (Fig. 3.2). Please note, PLT is the sum of touch time or work content (i.e. CT including NVA) plus waiting time of the WIP (waiting time is calculated as WIP in units times the following CT). Further, the information of takt rate (TR) imposed by the customer is an essential information which has to be compared to the process performance in order to check whether the exit rate (ER) of the process, i.e. the throughput, meets

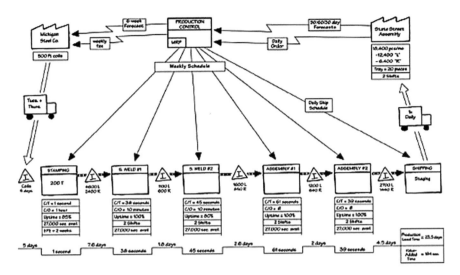

Fig. 3.2 VSM as proposed by LEI [2]

the TR or not. The TR is the number of pieces divided by a suitable time unit, e.g. 250.000 gearboxes per year which may be broken down to 50.000 units per week or 1.000 pieces per day (on a 5 days week basis) or 125 pieces per hour (on a 1 shift 8 h basis). It is not the intention of this book to transmit the basics and syntax of VSM; for that purpose we refer to the original literature [2], as well as regarding an extended treatment of the Lean metrics measuring the performance of the process we refer explicitly to [3]. The benefit of VSM have been largely discussed in industry journals [e.g. 4, 5].

In this section we try to evidence how to exploit the full potential of industrial VSM applied to transactional processes. Indeed, in transactional processes not the classic VSM is used but the so-called swim-lane representation of processes. Swim-lane process charts are similar to Makigami graphs showing the handover of transactions. Additionally, Makigami charts show also documents, database, and media used for communication; this is especially helpful in an office environment. In office processes the handover of entire files or single documents between different office departments causes waiting time of the transaction object. The swim-lane representation allows to highlighting these sources of waste.

This leads to enounce the

Principle of "Make extensive use of graphical swim-lane-type of process flow description to understand service performance generation".

However, the classic swim-lane representation is usually used to document processes in quality manuals or in descriptions of standard operating procedures (SOP) and therefore does not show the rich information content of a VSM. Also commercial software packages do not show useful representation of swim-lane VSMs. Nev-

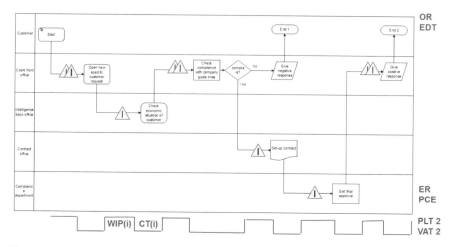

Fig. 3.3 Swim-lane process diagram enriched with process information (swim-lane VSM)

ertheless, it is imperatively necessary to add to the swim-lane diagram the VSM information, such as cycle times (CT) of the activities and especially WIP as well as the TR. Unfortunately, hardly any software is supporting such a swim-lane with VSM information. The WIP in transactional processes, of course, are not physical components or products but files and documents related to customer orders in the case of operations or internal orders according to the type of routine requested by the organization; the WIP allows to calculating PLT as well as PCE. We will call such a modified swim-lane, to distinguish it from Q-manuals or SOP documents, swim-lane VSM (Fig. 3.3). Due to the peculiarity of office workplaces and office employees we need to further distinguish the WIP. Indeed, employees are not executing only one task at their work place such as blue-collar operators on the shopfloor, they have to perform different tasks during a day. This makes it necessary to distinguish WIP in "primary operational WIP" (WIP generated by queuing of the main job type under scrutiny) and "other WIP" (WIP generated by remaining queuing due to the occupied white-collar employee executing another activity). We will simply double the WIP symbol to indicate such a multifunctional workplace. The waiting time may be split. We will come back to this in the Sect. 3.8.

Further, we have to define on which level we document the processes. In principle, this can be done on various level of abstraction, depending on what to put the focus. Therefore, we try to define the following activity levels for transactional processes (Fig. 3.4):

– Process: the process is the highest level of documentation and describes an end-to-end self-contained, self-sufficient transaction, e.g. an operation description involving different office departments constituting a value stream; this is the real domain of swim-lane VSM. Here the swim-lanes represent the departments and the activity boxes represent usually sequentially or parallel executed sub-routines (sub-processes), rather less the listing of single tasks, which indeed might be aggregated

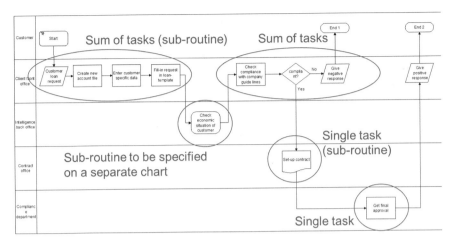

Fig. 3.4 Swim-lane of a process showing sub-routines and tasks

into a sub-routine. At any hand-over of documents between the departments WIP will materialize and has explicitly to be shown. This level of representation is very useful to right size the office departments for short PLT and adequate ER in order to comply with customers' expected delivery times (EDT) and order rates (OR). In service organization, instead of talking about ER it is more appropriate to talk about completion rate; nevertheless, we will use both indifferently but name it commonly ER.

− Sub-routine (or sub-process): the sub-routine is a self-contained sequence usually performed in the same office department by a team (or a single person), e.g. monthly account reporting. Here the activity boxes represent sequential or parallel tasks. The concept can be compared to a sub-routine in software development where variables are passed from the main-program to the sub-program, which returns afterwards an executable partial result to the main-program. Such a sub-routine might correspond to a work cell in industry as we will see in the Chap. 4. If the tasks of the sub-routine are performed by a single employee (integrated work organization), no WIP between the tasks should be shown, because executing one task after the other of a single job no WIP materializes (see Sect. 3.7). Several sub-routines together result in an end-to-end process. Here we can talk about sub-routine PLT, which we could also call routine lead-time (RLT).

− Task: the task is the minimum logic work content within a sub-routine (or a process) which should be performed by a single employee without interruption, e.g. enter data. The importance of being performed without interruption we will see in the Sect. 3.6. The addition of several tasks form a routine. Here we cannot talk about PLT; here we have to talk about cycle time (CT) of a single task. The CT_i is the time one employee at the operation i needs to perform one unit of an entity (file or document); the inverse of CT_i is ER_i and ER_i represents the number of units produced per time frame at this operation. Every operation (task) may have a different CT_i but it is advisable to balance the CT if different employees are involved as we will see in Sect. 3.7.

– Step: the step is here enounced only for completion reason because usually not of relevance for swim-lane VSM. We define step the smallest documentable and executable elementary process activity (part of a task) done by a single employee, e.g. enter name in a template file; the related task combining several steps would be enter customer data in a template file. This level of documentation is usually needed for industrial standard work such as methods-time-measurements (MTM) approach and will be discussed in the Sect. 3.3.

However, for the description of single employees' work, VSM is not apt. For that, standard operation procedure (SOP) in form of checklist suits best. The description of the single task can be delegated to the employee, knowing best what and how a task or a sub-routine has to be done. It is important, that such SOP exist in order to guarantee the reproducibility of the single tasks as well as for the entire process.

In association with a swim-lane VSM we can report the cycle times (CT) of the tasks or the PLT of routines (RLT) on a time operation/operator balance (TOB) chart (Fig. 3.5). This chart allows to performing a bottleneck and time trap analysis of a process or sub-process applying the Theory of Constraints (TOC) analysis [6]. We define here as bottleneck the task with the longest CT and therefore every process or sub-process has one and only one bottleneck [3]. Any CT, which exceeds the takt time (TT), is called a constraint; the presence of a constraint makes it impossible to meet the customer requirements of short PLT and OTD, as we will see. The bottleneck and time trap analysis is therefore an important topic in Lean because time traps are generating WIP and WIP causes delay in the PLT. Indeed, the bottleneck with CT_b determines the ER of the whole process or sub-routine (Theorem of Throughput, as we will see). In this book we avoid explicitly the mathematical dimension of Lean—for a comprehensive more mathematical discourse of this important topic we refer explicitly to [3]—although, we need a minimum to understand the achievement of on-time-delivery (OTD).

The Lean metrics summarizes the performance of a process or a sub-routine. Performance is intended here with short process lead time (PLT) and on-time-delivery (OTD) as well as, of course, meeting the required demand, i.e. matching the order rate (OR), by having the right capacity, i.e. sufficient staffing. We will substitute the concept of takt rate (TR) and the inverse notation of takt time (TT) used in industry because the TR assumes a regular demand with low variability whereas we will interpret OR as a wider concept with not defined variance. The inverse notation of OR, i.e. order time (OT), is the average inter-arrival time of orders. All the sizing of the office departments, i.e. the capacity in terms of full time equivalents (FTE) will be made in function of the mid-term E[OR] giving the structural load of the department (see Chap. 4). Overstaffing to the upper natural variability level is not recommended; recommended are flexible time contracts or in addition temporary limited hired personnel to meet the peaks. In certain cases this concept may be adapted to the contingent situation, e.g. call centers or tellers or seasonality, which need a variable staffing according to fast changing OR to guarantee a short and stable

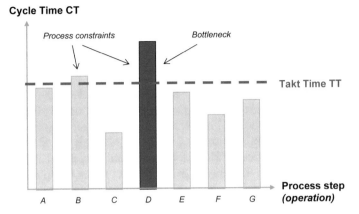

Fig. 3.5 Time Operation/Operator Balance Chart (TOB) with CT compared to TT showing bottleneck and constraints

PLT. The definition of bottleneck as the longest CT is synthesized in Eq. 3.1a. The ER of a process is given by the bottleneck of the process, or the sub-process, and can be expressed by Eq. 3.1b

$$\begin{cases} CT_b = \sup\{CT_i\} \\ ER = ER_b = \dfrac{1}{CT_b} \end{cases} \tag{3.1a, b}$$

Equation system 3.1 represents the definition of the bottleneck. We call the mathematical operator associated to the CT sup (or inf) and not max (or min), because of stressing the fact that we are not applying first order condition of a derivative function but we are searching the largest (or smallest) number of a defined set of numbers (i.e. CT). This property is so important that even a theorem has been enunciated [3] which is also valid for transactional processes

Theorem of Throughput (or Bottleneck Theorem):

Given is a sequence of production steps with each process step having a deterministic but different cycle time CT. The maximum throughput, i.e. the maximum exit rate ER, of a process is given by the slowest process step, i.e. the process step with the longest cycle time CT; this process step is called bottleneck.

First Corollary to the Theorem of Throughput (Corollary of Bottleneck Uniqueness)

Based on the definition of bottleneck, every process has one, and only one bottleneck.

Second Corollary to the Theorem of Throughput (Corollary of Bottleneck Time-invariance)

For a mono-product cell or transfer line, if no structural changes are made to the cell or line, the bottleneck is a time-invariant property of the process. Based on the intrinsic definition of bottleneck, only by applying changes to the present bottleneck by reducing CT or by mix change may generate a different, new bottleneck.

The critical importance of the bottleneck of the process determining the output of the whole process leads to the

Principle of "Concentrate the main attention always on the bottleneck because it limits directly throughput".

To observe OTD two requirements have to be fulfilled. In the case of a continuous order flow with low variance (SD[OR] = 0) we can talk of TR. First condition, the lowest ER_i of a sequence of i operations (i.e. the longest CT_i) has to be larger than the TR; this condition is formalized in Eq. (3.2a). This condition reflects the capacity aspect of the process or routine. Second condition, the PLT_Z, i.e. the time span from the entry point Z in Eq. 3.2b the process to the completion has to be shorter than the expected delivery time (EDT); this condition is shown in Eq. 3.2b. This condition reflects the speed aspect of execution. If condition 3.2a is not fulfilled, condition 3.2b will never be fulfilled. Both aspects of Eq. 3.2 form the necessary and sufficient conditions to observe OTD. Equation system 3.2 is valid if SD[TR] = 0 because no WIP with BWT (backlog waiting time) will form at the entry point Z.

$$\begin{cases} \forall i : \inf\{ER_i\} \geq TR \\ Z : PLT_Z \leq EDT \end{cases} \qquad (3.2a, b)$$

In the case that the TR shows high variability, as we have told earlier, we will substitute the notion of TR and rather talk about OR. In this case condition 3.2a becomes condition 3.3a where E[OR] means the expected value, i.e. the average, of the OR. A high variability (high standard deviation or variance) of OR, i.e. SD[OR] > 0, may lead to not observe anymore the condition expressed in Eq. 3.2a with the consequence of creating a WIP [3]. In this case it is advisable, in order to not congest the process, to stop and queue the orders at the entry point Z and to release the orders in the rate of completion, i.e. the ER. This leads to stabilize the WIP and therefor to have a constant PLT_Z. This means to apply the CONWIP or generic pull concept about which we will talk in Sects. 3.7 as well as 3.8. The orders at the entry point Z will form a waiting queue with the consequence of a backlog waiting time BWT_Z at the order entry point. In addition, this allows to prioritizing orders by changing the FIFO (also called FCFS first-come-first-served) queuing scheduling and release order in the case of preferential VIP customer treatment by "priority" LIFO (last-in-first-out) order release or HPFS (highest priority first served). Indeed, to observe EDT, in addition to the PLT also the backlog waiting time BWT has to be taken into consideration. The BWT and PLT constitute the customer visible time CVT. If Eq. 3.3b might not be respected one has to increase first capacity, if also Eq. 3.3a might not be respected; this means acting on the left side of Eq. 3.3a. If condition 3.3a is respected, one has to change entry point Z to observe OTD. Equation system 3.3 is also valid to supply JIT of supermarkets where BWT_Z then represents the backlog of Heijunka box.

$$for : SD[OR] > 0$$

$$\begin{cases} \forall i : \inf\{ER_i\} > E[OR] \\ Z : BWT_Z + PLT_Z \leq EDT \end{cases} \tag{3.3a, b}$$

These important conditions/equations for OTD have also been translated for production theory into a theorem. This shows once more that Lean is more than just Muda elimination; for further information consult [3].

Theorem of General Production Requirements (or OTD Theorem):

The necessary and sufficient conditions to supply a customer with OTD, i.e. with the right quantity at the right time, is that first the capacity requirement and second the lead time requirement have to be satisfied simultaneously, independent of the applied transfer principle, i.e. SPF or B&Q. The capacity requirement is given by the Corollary of Weak WIP Stationarity and the lead time requirement necessitates that MLT plus BWT is shorter than EDT

where the PLT (i.e. the time one piece needs to transit the process) has been substituted in manufacturing by the MLT (manufacturing lead time), i.e. the time until the first piece of the batch enters the process and the last piece of the batch exits the process.

It exist a branch of mathematics dealing with queuing theory. Queuing theory is an important topic for batch operated push production systems, less for TR controlled pull production systems (I prefer to say TR controlled and not to say in this context TR synchronized because due to the Kanban supermarkets, semantically speaking, we have asynchronous governed JIT processes [3]). For our purpose, we can take the simplest form of queuing expressed by Little's Law in Eq. 3.4.

$$PLT \approx \frac{WIP(t)}{ER} = WIP(t) \cdot CT_b \tag{3.4}$$

We can calculate the PLT of a process or routine adding the WIP in front of the bottleneck and divide it by the ER. The result gives the time one item (file or document) needs approximately to transit the process. This formula is generally taught by consultants, widespread used and universally applied. However, if you know the bottleneck, a better approximation is given by Eq. 3.5

$$PLT = \sum_{i=1}^{n} CT_i + \sum_{i=1}^{b} WIP_i \cdot CT_b \tag{3.5}$$

where n > b. For the correct calculation, please consult [3], but for a first approximation Eq. 3.4 is sufficient. If we consider the value-add time (VAT) and non-value add (NVA) concept to calculate the performance of a process or routine, it is the intention to minimize, or better to eliminate, the NVA part. Equation 3.6 gives the process cycle efficiency ratio (PCE) aka Lean indicator.

$$PCE = \frac{VAT}{PLT} \tag{3.6}$$

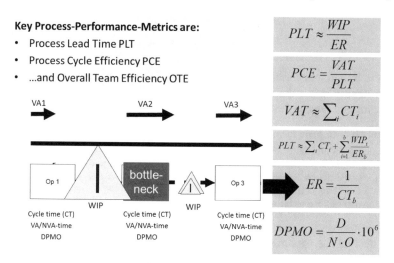

Key Process-Performance-Metrics are:
- Process Lead Time PLT
- Process Cycle Efficiency PCE
- ...and Overall Team Efficiency OTE

$$PLT \approx \frac{WIP}{ER}$$

$$PCE = \frac{VAT}{PLT}$$

$$VAT \approx \sum_i CT_i$$

$$PLT \approx \sum_i CT_i + \sum_{i=1}^{b} \frac{WIP_i}{ER_b}$$

$$ER = \frac{1}{CT_b}$$

$$DPMO = \frac{D}{N \cdot O} \cdot 10^6$$

Fig. 3.6 Lean main metrics overview

This ratio, usually expressed in percentage (%), shows how efficient in terms of value-add tasks or value-add steps consists the PLT; a low PCE indicates a lot of NVA what Japanese call Muda (waste) not adding value in the eyes of the customers. According to Taiichi Ohno everything for what the customer would not pay is waste. There is another category of activity which sometimes is called in different ways (business non-value add time, ancillary time) but which we will call process required time (PRT); these tasks, or better steps, are required for the routine or task to be performed. They are needed but do not add value, e.g. fill-in data in a template. Different from industry, where we distinguish only VA and NVA, it is advisable to use in transactional environment the category of PRT and not to call it NVA if it is not really NVA, because many tasks in the office belong to this category; otherwise employees may be shocked about the low VA work they contribute. The Lean metrics is summarized in Fig. 3.6.

These are the basic Lean metrics to measure the performance of a process or sub-routine; for a deep dive in process performance we refer to [3]. Summarizing, the Lean office metrics does not differ from industry metrics and mandatorily needs to be computed in order to improve the performance of processes and sub-routines. In the next sections we will learn about other metrics e.g. regarding team efficiency in Sect. 3.4 as well as quality in terms of DPMO in Sect. 3.5.

3.2 5S, Mieruka and Organization

Should the picture of the latest family vacation on the employee's desk be tolerated in the office? In a shopfloor 5S environment it is not—and the pinup girls in the locker cabinets have meanwhile disappeared, because it is neither useful nor necessary for performing the task and may even distract the operator. Now, why should a white-collar employee be treated differently than a blue-collar worker? Everybody may answer this question by himself.

One is for sure—there is no Lean without 5S, today sometimes also called 6S in industrial environment, adding the safety aspect to the work place. The emblematic identification of 5S with Lean is why many Lean initiatives in Europe begin with 5S. But frankly speaking, with 5S and shopfloor management a company has not yet introduced Lean, as sometimes puppy-like believed, and for sure, it will not save or earn a lot of money and gain competitive advantage! Due to the rich literature regarding 5S we will limit the discourse on this topic to some key advising comments and refer e.g. to [7, 8]. But what is 5S? In a nutshell:

> 5S is an efficient workplace organizational approach, in order to have a Lean workplace with all needed tools, and only the needed ones, as close as possible to the operator avoiding movement waste of the operator.

This facilitates the operator in doing his job and may present the desired side effect to increase productivity. The 5S approach is divided into five steps:

- Sort (buzzword "red tagging")
- Set in order or straighten (buzzword "shadow board")
- Shine (buzzword "white floor")
- Standardize (buzzword "check list")
- Sustain (buzzword "endorsement").

Every step is necessary; do not skip one. If a company is introducing 5S it has to be done seriously with management commitment. This is necessary because hardly employee will endorse the 5S approach from the beginning if not exemplarily exercised by the principal. A not sustained 5S initiative is often immediately related to a Lean failure; no, not Lean is failing—the failing is attributable to a bad management! If a company embarks into the journey to introduce 5S in the office it is advisable to appoint a 5S responsible, at least at the beginning. Nothing is more embarrassing, for not saying ridiculous, if "shadow boarded" white board markers in a 5S-organized meeting room are missing or not writing. Further, the set in order has to be standardized uniformly throughout all office working places in order that everyone sitting at a new place feels immediately comfortable (confront car models in automotive); this is important for shared desks (if a small personalized zone for every employee is granted, this is left to the discretion of company management). However, it has to be carefully evaluated if such a clinical office layout will enjoy employees in Western companies and if it is worth to be forcedly introduced causing a hostile mood of employees. This approach might be correct for shared desk but not for individually assigned work places. On the other hand, todays increased competition is reducing the degrees of freedom for employees, and if OPEX is the aim, the way how to proceed is given.

The layout of the workplace can be standardized; this is recommended for international companies. The workplace could be composed of two different zones: a standard zone with general working tools and a specific zone with workplace dedicated working tools. However, often the layout is contingent, such as e.g. in surgery laboratories, where each medical doctor has its own technique; and he will dictate the tools layout. Would you risk to embarking into a deleterious discussion on these items? It is better to lose a battle and to win the war.

An important concept on the Toyota shopfloor level is Mieruka, which can be freely interpreted as visual management. Visual management is very important because human beings acquire information best by graphical vision. In a nutshell, four types of Mieruka exist:

- Identification (e.g. product barcode)
- Informative (e.g. actual production level)
- Instructional (e.g. SOP)
- Planning (e.g. time scheduling)

and three rules have to be observed:

- Easy understandable (e.g. colors)
- Easy visible (e.g. Andon)
- Easy changeable (e.g. white board markers and not PC).

Informative and planning Mieruka might be displayed on the Kaizen board (Kamishibai or master board, or shopfloor board, however it is called). Identification Mieruka is product related and instructional Mieruka are task and workplace related. These concepts can be transferred as is to the office environment. Identification can be facilitated by colored classifiers, instructional are the standard operating procedures (SOP) as well as the quality manual instructions. Informative Mieruka usually are performance indicators related. For further information consult e.g. [9].We can anticipate, due to the fact that WIP and workload is not visible it has to be made visible by flags or Andon lights. In the Sects. 5.1 and 5.2 of Chap. 5 we will come back to Mieruka.

The office 5S tool leads to the

Principle of "Strive for uniform layout of the office desk workplace"

facilitating the working conditions for every employee setting the conditions for a productive working environment.

3.3   Normalized Work and Reproducibility

At first, we have to correct a widespread, consultant originated, misinterpretation of standardized work. The primary aim of standardized work is not to having SOPs defined nor to have a baseline for the continuous improvement PDCA approach (Deming's Plan-Do-Check-Act cycle); this may be a derived and welcome side benefit. However, standardized work is fundamental to implement single piece flow (SPF) in work cells. We can define standardized work as follows [10]:

Standardized work is the optimal combination of operators, machines, and material to ensure that a task is completed the same way every time with minimum waste to consistently meeting TT. Standardized work is made up of the elements TT, work content (WC), work sequence, standard WIP, and adequate staffing. Standardized work should be defined, maintained, and improved by operators and supervisors.

Indeed, in a Heijunka box scheduled and takted SPF mixed model cell, i.e. a work cell dedicated to multiple similar but different products, it is of utmost importance that PLT and ER meet the demand to replenish timely the Kanban-controlled supermarket in order to avoid a stock-out. For that, the operators' work has to be repeatable and reproducible. Useful tools in this context are Process Capacity Tables and Standardized Work Combination Tables for which we refer e.g. to [10, 11]. We will not enter here in the large and interesting topic of industrial cell design, with additional concepts of workstation (WTT) and cell turnover time (CTT), for which we refer e.g. to [3] but try to transpose part of the technique to the office context.

In industrial context, the stopwatch, i.e. the chronometer, is an essential tool to implement standardized work. Measuring the CT of the tasks, even of each single step, is mandatory. This is not only for improving the baseline, but also to be able to plan and implement the cell. For that, the work content is defined and executed by the best operator (let us be realistic, for the beginning the second best is more appropriate) which rhythm has to be attained by all other operators. In Sect. 3.1 we defined the logic activity entities as

- Step
- Task
- Sub-routine
- Main-process.

Standardized work in industry is done on level "step". Due to the intrinsic difference of a transaction compared to a product, in a transaction every new order is different even in the same product category approaching a real batch-size 1 for every transaction. Due to this characteristic, the activity entities "step" and "task" may present variable content, and therefore variable CT, e.g. the granting of a bank credit may depend from contingent factors for which more time is needed. It is therefore recommendable, at least at the beginning, to leave a certain amount of discretionality to the employee defining an upper limit to execute a task or even to define the maximum acceptable PLT for a defined routine, upper limit of course to be challenged by the supervisor. Indeed, although also e.g. credit institutes talk about products, the product content is highly variable and differs from each single order. It makes therefore more sense to talk of the **product in service industries as a parametric template with defined structure but with variable content**. With each order, the content has to be filled in the template structure to be then processed. This is the reason for what we prefer not to talk about standardized work in the office environment, implying strict observation of time for every step (because executed in high performance cells), but we prefer to talk of normalized work, allowing a certain discretionality in delivering the routine. Let us **define normalized work as a "contingent work standard" to complete a task within a certain timespan according to a defined**

Routine:		Product type:								Observer/Date:				
No.	Task	1	2	3	4	5	6	7	8	9	10	avg Time	Employee	Comments
	Time for 1 Cycle													

Fig. 3.7 Process time observation chart

SOP. It is preferable to talk about normalized work (in office environment) instead of standardized work (industry) because of non-deterministic transaction content compared to the deterministic work content of an industrial product. Therefore, the focus should be switched from cycle time of a task to transaction time of a sub-process because the non-deterministic content leads to a "normalized execution" per transaction. Nevertheless, also for transaction-based order transformation, the stopwatch becomes important and the time measurements can be recorded using a process time observation chart (Fig. 3.7). The repetitive records help to stratify the products in categories, e.g. simple, medium, complex, taking different time to be processed (see below).

Please note, the sum of the tasks corresponds to the work content (WC). If the tasks are executed without interruption, i.e. no waiting time between the tasks, then the WC corresponds also to the routine PLT. If during the routine a task needs input from another department, waiting time is the consequence and it may build-up a WIP. In that case the PLT of the routine has to take this waiting time into consideration; an alternative is to split the routine into part one and part two separated by the WIP. This is recommendable if the waiting time is long in respect to the WC or the two parts operate at a different pace; this allows to switching to another activity in the meantime.

Due to the intrinsic difference between manufacturing and office—remember, due to the parametric content of a transactional order it is difficult to prescribe the time of a sequence of single steps within a task in a detailed SOP—not the exact temporal sequence has to be prescribed but in which time frame the result has to be delivered. Of course, if the times may vary, the PLT of the procedure, sub-routine or process, will vary but should not vary too much, and not between employees. To limit discretionality of the employee it is possible to define classes within a product type performed in the same routine, e.g. simple transactions, transactions of medium time absorption, and complex transactions. This of course leads to variable PLT of the same routine but the variability is not due to non-repeatability or non-reproducibility of the order but due to the variable content of the same type of orders. Indeed, as we have already seen in Chap. 2, **due to the fact that the performance of a service is determined "parametrically" through the uniqueness of each single order specification we cannot talk of an "identical repeatability" (and reproducibility) of the orders, however, we can at best talk of a "formal repeatability" (and reproducibility) that relates to the repetition of the process but not to the content**. And exactly this lack of "identical repeatability" and "identical reproducibility", in most of the processes, will lead us towards a different modeling of the office desk job compared to manufacturing job: one that is more form-like (functional and relational instead of procedural), but also defined by the target instead of a deterministic algorithm. The SOP has therefore rather the character of a checklist.

At the begin of this section we said that standardized work is not synonym with standard operating procedure (SOP); indeed, SOP might be an input to define standardized work. However, SOP are also important on office-floor in order to guarantee the formal reproducibility of processes by multi-skilled employees. SOP have to be easy readable making also use of checklist; these SOP help also novice employees to get fast familiar with the reproducible handling of transactions. Due to the fact that transactions are often performed by division of labor involving different employees necessitating meetings, SOP how to conduct or participate at meetings are of utmost importance to make the outcome most effective. Although this topic is very important we do not enter further into this discussion.

In order to approach the "formal repeatability" to an "identical repeatability" (as well as reproducibility of course) it is necessary to define a within-product stratification, which corresponds to a product mix, which allows to plan better the load of a team, i.e. the office cell, and to control the efficiency of the team (see next section). For the staffing of a cell we refer to [3] and to Chap. 4. Although it might feel strange to apply a time-bound methodic working in the office world, the increased global competition forces management to apply industry-proven productivity increasing methods also for the transactions of the office environment, independent of front or back office. Nevertheless, the applicability of strict time standards due to the variability of order content is difficult, as we have just seen. It is therefore important to specify upper specification limits (USL) for the CT, concept derived from Six Sigma capability theory, not only for each product but also for each stratification category within each product and to supervise the stability of the process with Six Sigma control charts [e.g. 12, 13].

The normalized work tool leads to the

Principle of "Observe the repeatability and reproducibility of a task or a routine for every employee"

in order to meet OR and EDT.

Within the context of a mixed-product cell, we have also to see the performance of the cell in terms of ER. The ER of an office cell will change in function of the mix due to non-deterministic parametric content of the order template reflected by the product/service stratification. Normalized work can help to limit the variability of ER and therefore to plan better the load of the cell. The required average E[ER] has also in the short term to be larger than the average order rate E[OR].

The variability of content, i.e. mix, may false the performance in terms of ER. Instead of talking of ER, i.e. the completion rate expressed in number of transactions per time unit, due to the variable content of the same product and therefore the inherent variable work content (WC) it is advisable to build standardized stratified classes (small, medium, large product content) and to assign **standard WC units** to the categories. This allows to measuring the performance in terms of **standard ER units** having a better comparison of performance. The prerequisite is to define upfront the estimated WC of each class and to attribute standard WC units to each order. We leave here apart the difficulty to define and measure, or better the time necessary to measure, the standardized ER at the end of the day.

3.4 TEM (OTE) and Efficiency

Total productive maintenance (TPM) is not an explicitly mentioned tool in Liker's version of the TPS house model. Nevertheless, for me it has the same importance such as the Jidoka pillar. Indeed, within a Heijunka pitch-scheduled and takted SPF, if a single machine or equipment fails, the whole production line is down, therefore the whole production stops and no piece exits the line anymore (we leave aside the consideration of identical backup cells). This is the Achilles' heel of SPF whereas in Western B&Q production systems generating WIP, the WIP separates the single workstations. Indeed, in the case of a minor hazard leading to an equipment failure, the WIP constitutes a sort of operational buffer allowing production to continue. This means, TPM is implemented in the TPS to assure high uptime (or to minimize downtime) in order to guarantee the required daily output. TPM is today a widespread topic originally divulged by Nakajima [14]. However, it has to be mentioned, based on own lived experience as a consultant, that in Western manufacturing companies TPM is often introduced on the most expensive equipment to minimize downtime (and reduce NVA and increasing potential VAT) with the credo that the most expensive equipment has always and continuously to produce parts. In the case the costly equipment is not the bottleneck, such a decision of the production manager reveals his ignorance about bottleneck TOC analysis. Therefore, TPM is primarily introduced to guarantee the required throughput. To do so, TPM is a comprehensive approach to assure operability of the whole equipment based on:

– Maintenance prevention (avoiding maintenance)
– Predictive maintenance (forecasting failure)
– Corrective maintenance (reducing need)
– Preventive maintenance (anticipating failure)
– Autonomous maintenance (regular care).

Exactly the last one is in the operator's own competence and can be implemented easily; the delegation of responsibility about the equipment increases his estimation to

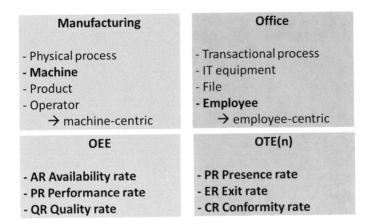

Fig. 3.8 Comparison between TPM (OEE) and TEM (OTE)

be a valuable part of a system. Together with TPM an important indicator is associated to each tracked machine, called the overall equipment effectiveness (OEE). The OEE is composed of three elements (Eq. 3.7), the availability rate (AR), the performance rate (PR), and the quality rate (QR); it is one of the most important KPI allowing to identify precisely the area of problem and where to intervene.

$$OEE = AR \cdot PR \cdot QR \tag{3.7}$$

Now how to transfer the concept of TPM and OEE to the office environment, which has no production equipment? In the transactional industry the IT infrastructure is pivotal and therefore of utmost importance for which an AR and PR may be computed. However, in the service industry the employees play a fundamental role in the organization, which performance at the end determines the productivity. Work force also represents a considerable cost weight in the P&L statement to which management pay close attention. Let us therefore talk in the context of Lean office management about Team Efficiency Management (TEM) and calculate the related Overall Team Efficiency (OTE) indicator. Please note, we call it expressly efficiency and not efficacy indicator. Figure 3.8 shows the parallelism between the manufacturing industry and service industry of the performance management regarding on the one-side machines and on the other side employees.

Whereas the OEE performance indicator does explicitly not consider the operator—the central attention focus on the equipment—in the service interpretation of the OEE, the new defined OTE(n) is focused only on employees and is also composed of three elements: the presence rate (PR), the exit rate (ER), and the conformity rate (CR). We prefer to talk of CR instead of QR in an office environment, in order to transmit a dedicated office feeling of the concepts but frankly speaking, there is no difference between conformity and quality. The OTE(n) is a team performance

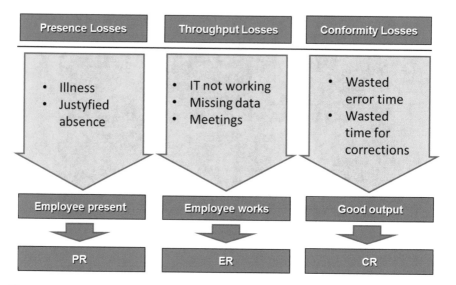

Fig. 3.9 Overall Team Efficiency (OTE) components: presence, exit, conformity rate

indicator and the (n) indicates how many FTE are composing the team to which the measure is extended. The formalization of OTE is shown in Eq. 3.8.

$$OTE(n) = PR \cdot ER \cdot CR \tag{3.8}$$

It has explicitly to be noted that related to the OEE exists the problem of how to define the baseline; the different interpretation by each company lacks of homogeneity, which therefore does not allow to comparing OEE between different companies. This is not an issue because the OEE is primarily an internal trend indicator, which serves management to act and is less considered to be a benchmark indicator. In Fig. 3.9 we see from what the different ratios are influenced.

The PR takes the presence loss of the employee into account, such as illness and justified absence; the PR reflects the non-presence at work of the team. The ER considers the performance of the team and takes into account losses due to IT standstill not allowing to working, reduced activity due to missing data, absence of the employee due to meetings; please note, the person may be present but not being able to work. The CR accounts for the quality of work making conforming products to specification; it considers the time spend making errors or the eventual time to make the corrections (taken alternatively). The OTE(n) reflects the efficiency for a perfect output; the output is a perfect delivered service. This key performance indicator (KPI) is perhaps the most important indicator to manage the performance and the cost of a team because it allows to intervening directly on the presence, throughput, or quality of work. The problem will be to gather the related data. We

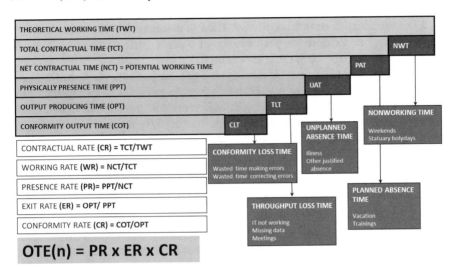

Fig. 3.10 Consistent scheme to compute OTE(n)

leave here apart how to gather objectively this waste; employee may also be reluctant to show their performance.

In Fig. 3.10 we propose a consistent scheme for the OTE(n) calculation. The consistency is enhanced by using only time units to measure efficiency performance. Due to the fact, that in many European working contexts it is not allowed to measure the performance of a single operator, we extend the calculation at a team or a department composed of n FTE. The timeframe has therefore to be multiplied by the number of FTE composing the cell or department.

In the context of Lean office we will not talk of downtime, which is machine related, but rather of non-activity. The non-activity might be planned or not; the unplanned reduces available operating time and the planned reduces directly the potential output. This non-activity corresponds to "unplanned absence time" and "throughput loss time" of the scheme shown in Fig. 3.10 giving the "output producing time" which may be compared to a net operating time of industry.

We limit our indicator to OTE(n) because its single ratios of which it is composed can be influenced by management; however, the other partial ratios indicated in Fig. 3.10, such as WR and CR, cannot be influenced and are forcedly given according to country labor laws.

Please note, we put here the emphasis on efficiency and not efficacy. Doing the right things (effectiveness), is triggered at the prioritization level, which will be treated in Sect. 3.8. Indeed, efficiency normally is naturally built-in in industrial processes but not yet in office processes.

The TEM leads to the

Principle of "Strive-for-team-efficiency in order to maximize productivity".

It has explicitly to be mentioned, that TEM with related OTE indicator is a very important tool to measure the performance of a team, performance which can be translated in terms of money, i.e. the cost associated to the performance of transforming input into output. However, if employees like their performance to be measured in such a way is doubtful, especially having experienced cost cutting restructuring programs. Indeed, Lean has often introduced in service industries not to reduce WIP-captured liquidity and to speed-up processes, but simply to reduce cost; cost cutting in service industries is always targeted to reduce the number of employees.

3.5 Poka Yoke and First Time Right

Processes have not only to be fast and the deliveries have to be on time; also delivering the right quality is essential, i.e. characteristics conform to specification. According to the SPQR axiom it represents even the central element of the business system. The voice of the customer (VOC) gives the driving behavior to assure quality. The cost of poor quality (COPQ) is of main interest within Toyota. The later a defect is detected the more it costs. Therefore, as soon as a defect is detected (out of control signal), the part is taken immediately out of production; never a defective part should be passed to the next workstation. Usually, at the same time, the line is stopped and the cause of the defect is investigated. Corrective actions are implemented and monitored based on Deming's PDCA cycle (plan, do, check, act). These were the origins of the Kaizen continuous improvement culture. We will come back to this topic in the Sects. 5.1 and 5.2 of Chap. 5. Within the TPS, Jidoka—aka autonomation, which means intelligent automation—has the objective to assure 100% quality of produced parts. Poka Yoke is a technique to implement Jidoka [15]. The SPF is a precondition for 100% quality, being each single piece automatically checked. Because human beings are making errors and therefore they are not reliable, the quality checks should not be performed by the operators but through technical devices. Poka Yoke is a tool within the Jidoka quality concept, which means "fool proof". Indeed, Toyota invented the in-line quality control opposed to the western final quality check. Japanese companies produce quality—Western companies check quality. This sounds heretical, but process quality is not a European strength but of American (e.g. Deming) and Japanese (e.g. Taguchi) origin. A definition of Poka Yoka might be [10]:

> Poka Yoke is any device which avoids the production of an error or which allows the detection of an error. Therefore, we can distinguish mainly two classes of Poka Yoke types: error preventive types and defects detection types. The detection type may be further split into controlling approach (stops the process) and warning approach (only signals a non-conformity to alert the operator to intervene).

Now, how can this concept of autonomation be transferred to the office environment? It has already largely been introduced, e.g. in the IT. Often IT-based applications have mandatory data entry fields. If a process still is not supported by IT, checklist may help to avoid errors. Checklists can be considered to be a sort of Poka Yoke, however a weak one because it does not prevent forcedly the occurrence of defects. In IT-based application, however, if certain entry fields are not filled-in, the application will not continue, or if an identification number shows a wrong syntax, it is detected and refused. The problem consists rather in the manually filled-in application sheets where human errors might be overseen. To show the quality level of first-time-right we have to measure it. For that we will introduce two quality measures which suit perfectly also the office environment; these KPI are especially suited for attribute data. The first one is Defective Parts per Million (DPPM) often simply named ppm shown in Eq. 3.9

$$DPPM = \frac{d}{N} \cdot 10^6 [\text{ppm}] \tag{3.9}$$

where d denotes the number of defective parts and N is the sample size. Quality KPIs are usually reported to 1 million parts. This KPI follows the binomial logic of false or true, passed or not passed. This quality metric is widely used. The second one is Defects per Million Opportunities (DPMO) shown in Eq. 3.10

$$DPMO = \frac{D}{N \cdot O} \cdot 10^6 [\text{dpmo}] \tag{3.10}$$

where D denotes the number of defects, N is again the sample size, and O the number of error opportunities. This KPI follows the Poissonian logic, where D/N corresponds to defects per unit (DPU) also known as lambda. The DPMO measure is widely unknown, or better not applied, in European companies. In Eq. 3.9 it does not matter if one or more defects are on the same piece; it is sufficient that a piece presents at least one defect to be refused. However, the average probability to make an error on a piece is given by DPU and, remember, a process may produce defects and not defective pieces, which is a consequence of potential defects. A quality management system has primarily to operate with Eq. 3.10 to improve the quality level of their products, customer of course will apply Eq. 3.9; with Eq. 3.9 one can measure quality but not improve it. This is an additional sign of European ignorance regarding theoretical quality management.

A very useful tool to improve quality is the Pareto chart. The Pareto chart is a sequence-ordered type of a sort of histogram with attribute abscissa. The ordinate gives the frequency of each attributive characteristic. The characteristics are observed errors. Due to the frequency-ordered sequence of characteristics the chart allows to concentrate on the "many vital" errors leaving aside, at first instance, the "few trivial" leading to the so-called "80-20" rule (Fig. 3.11).

Now what are the opportunities in an office to make errors? Let us take an example. If a shipping bill contains the address, the product identification, the number of pieces, and the price, then potentially all four partial information of the shipping bill may experience a wrong input, so there are 4 opportunities O to make an error. If during a day on a sample N of 200 shipping bills are found 15 defects (D) on 10 bills (d), i.e. we discovered 10 defective bills out of a sample of 200 totaling 15 defects,

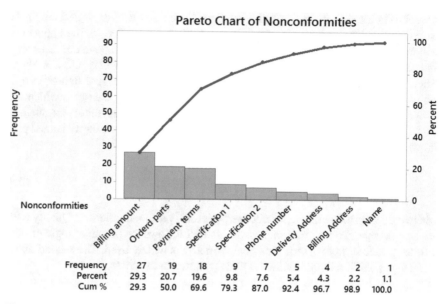

	Billing amount	Orderd parts	Payment terms	Specification 1	Specification 2	Phone number	Delivery Address	Billing Address	Name
Frequency	27	19	18	9	7	5	4	2	1
Percent	29.3	20.7	19.6	9.8	7.6	5.4	4.3	2.2	1.1
Cum %	29.3	50.0	69.6	79.3	87.0	92.4	96.7	98.9	100.0

Fig. 3.11 Pareto chart showing the frequency of defects

then both quality KPIs can be easily computed (18,750 dpmo, 50,000 ppm). Pay attention, consider only real error opportunities, which can also happen, in order not to "improve" artificially the DPMO quality measure.

In the case we deal with continuous data, e.g. such as PLT compared to an USL (upper specification level), then we can use the Cpk measure shown in Eq. 3.11

$$Cp_k = \frac{USL - \bar{x}}{3\hat{\sigma}} \tag{3.11}$$

and e.g. for the observation of PLT, Eq. 3.11 becomes

$$Cp_k = \frac{USL - E[PLT]}{3 \cdot SD[PLT]}$$

Conforming to industry standard the Cpk should be >1.33 to say that the process is capable. If the USL is set to be e.g. 5 days and the process shows an average PLT of 4.5 days with a standard deviation of 0.2 days, then the Cpk results to be 0.83, i.e. by far too small to be compliant with the quality standard.

The occurrence of an error is a sign that the process is out of control or even out of specification. As soon as a defect is detected, the Japanese have been setting-up a Kaizen team to investigate the cause and to implement a corrective action leading to the root cause and corrective action (RCCA) approach. In that way, case after case, all potential error causes have been eliminated, so that the error will not occur any more. You may smiling, but in European manufacturing culture instead of searching for the causes, the operator would have been blamed and invited to work harder. Simplified, that is the reason why Japanese quality has become famous. Now, we propose the following eight-step approach each time when a defect is detected [10]:

– Describe the defect
– Describe the circumstances where the defect has been detected
– Identify the location where the error has been made
– Analyze the SOP at the location where the defect has been produced
– Identify errors in or deviation from the SOP
– Investigate and analyze the root cause for each deviation
– Brainstorm ideas to eliminate or detect the deviation early
– Create, test, validate, and implement Poka Yoke.

It is said that Shigeo Shingo would have said that only the limitation of human imagination can impede to find a suitable Poka Yoke; there always exists a Poka Yoke, a good or worse one. An additional very helpful tool and approach for the RCCA procedure is the Failure Mode and Effect Analysis (FMEA) but also the cause-effect C/E matrix [10].

Since the last 30 years, the Six Sigma quality management approach has conquered industry. This statistical process control (SPC) quality management system based on control charts and process capability aims at 3.4 dpmo. Recently this quality management approach has been merged with Lean, so that often we hear talking of Lean Six Sigma (LSS) [16]. Control charts (Fig. 3.12) help to monitor the stability of the processes. Stable processes are a precondition to calculate the process capability. Figure 3.12 shows the individual moving range (ImR) control chart of the PLT of 22 order inquiries. As soon as an out-of-control (OOC) signal is detected (the monitored KPI goes beyond the UCL upper or LCL lower control line) the root cause has to be identified and eliminated. In Fig. 3.12 this is the case of the 16th observation.

If we apply Eq. 3.11 with an USL of 7 days, this is the VOC specifying the expected delivery time (EDT) of the order, we get a Cpk = 0.13 which is fairly low; we say, this process is not capable. Indeed, 23% of inquiries are above the specified EDT of maximum 7 days (see Fig. 3.13).

Nota Bene: due to the occurrence of an OOC signal, theoretically the Cpk may not allowed to be calculated because the OOC observation does not any more guarantee the correctness of the capability calculation. In addition, the non-normal shaped density distribution of PLT would not allow to using Eq. 3.11 because based on underlying normal-distributed data. Consult for further information on control charting and capability analysis e.g. [12, 13].
This leads directly to the

Principle of "Measure your performance to be able to improve performance".

In parallel, very popular became the LSS DMAIC (define, measure, analyze, improve, control) problem solving approach. This DMAIC approach has often been applied in OPEX initiatives; however, a problem solving approach cannot change the culture of a company although it is at the beginning of an OPEX journey a very effective approach to improve fast and significantly performance.

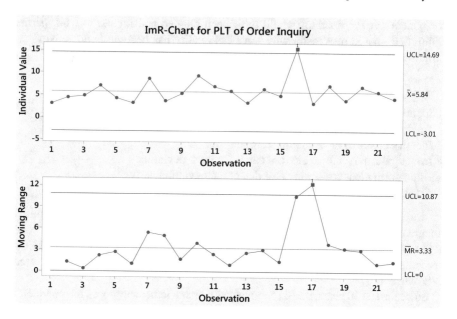

Fig. 3.12 SPC implemented with control charting to monitor specified EDT

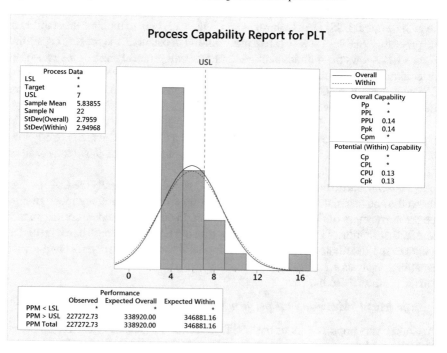

Fig. 3.13 Capability analysis regarding PLT of order inquiry

The Poka Yoke leads to the

Principle of "Assure VOC conformity by first time right"

in order to increase process build-in quality. We can conclude: monitor quality and finally check quality is not enough, we have to produce process build-in quality and for that, Poka-yoke technique is a valid tool.

3.6 JOSO and Flexibility

Usually, a manufacturing cell is not dedicated to a single product; similar products, needing the same process sequence, are grouped together in the same cell and scheduled alternatively. However, different products on the same manufacturing line consume time for changeover, e.g. changing the dies of the equipment and the process settings of the machines, loosing valuable potential production time. For this purpose the changeover time (C/O) must be reduced. We can define C/O as follows:

> Change-over time in industry is the length of time from the last good product of a production run A to the first good product B of the next production run [10]. It comprises search for dies, mechanical set-up, adjustments, and run-in phase.

Western companies preferred in the past to have dedicated manufacturing cells in order to avoid C/O with the credo to increase productivity by eliminating product set-ups. The result might lead to increase capital expenditures and not fully used production capacity. In addition, a mixed-product cell is less vulnerable to order rate variation. Indeed, the average standard deviation of the cell is smaller than the standard deviation of the population and is related to the number of products manufactured in the cell following the central theorem of statistics

$$\sigma_{\bar{x}} = \frac{\sigma}{\sqrt{n}}$$

where n is the number of different products manufactured in the cell.

However, reduced C/O allows to reduce the batch size, a reduced batch size generates less WIP, less WIP mean shorter PLT. Toyota (with the help of Shigeo Shingo) developed for this purpose a four step rapid set-up (or changeover) method known with SMED (single minute exchange of die [17]). The four steps of SMED are:

– Observe, list, and measure present sequence of C/O tasks
– Categorize tasks in internal and external and transfer internally executed tasks to external if possible
– Optimize remaining internal task and apply 5S
– Eliminate all adjustments.

This technique is also applied in European manufacturing plants but with the intention to reduce NVA and to increase VA production time. Remember, SMED

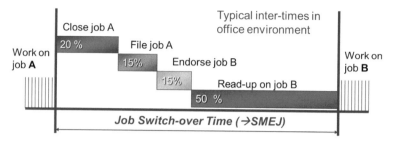

Fig. 3.14 Definition of JOSO (SMEJ)

is performed to allow a mixed product cell production modality. Toyota shortens C/O time to increase flexibility of a mixed product manufacturing work cell. We see here another testimonial, that European companies have not understood the real significance of Lean—Lean is not only Muda elimination but presently the most modern and performant manufacturing theory [3].

Now, how to apply this concept in transactional service companies? On a first instance, one could say SMED is not applicable in the office, because having no machines it makes no sense. However, if performance in terms of employee productivity (OTE) becomes critical, it will be mandatory to think also for the office environment in this direction. Instead of talking about C/O, we will talk about Job-optimized switch-over time (JOSO). We can define the office job switchover as follows:

> The JOSO time in the office is the NVA-time lasting from closing and filing of the last job (time included) until the beginning VA-time of working productively on the new job. The job might be a task or more likely a sub-routine. The new sub-routine might be one of similar type or one of another type of activity (in non-specialized multi-routine department).

This situation is shown for a job in Fig. 3.14 and is called Job Switch-over Time. The JOSO is performed by applying the technique of Single Minute Exchange of Job (SMEJ).

Here the single phases of JOSO comprise:

– Closing the old job means complete all necessary content-related tasks
– Filing the old job means verifying identification and store it according to 5S-Mieruka conformity to facilitate retrieval
– Endorsing a new job means taking the new job in charge and create the file
– Read-up on the new job means getting familiar with the content.

It is assumed that the jobs are processed in a FCFS way. Different from shopfloor, there is an additional type of set-up, an improper one, not at the end of the job but during the productive phase of the job execution due to a disturbing event such as a telephone call or answering to a supervisor's question, or a very urgent request, causing an interruption of the present job. Employees are reacting differently in the case of disturbing events; two types of people exist:

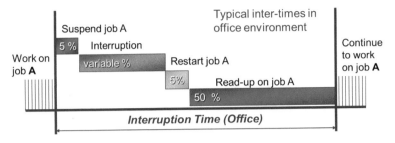

Fig. 3.15 Consequences of a job interruption

– flexible-minded people and
– narrow-minded people.

Flexible-minded employees are multitasking able and can handle two parallel tasks. The parallelism is of course virtually and can be compared with the CPU cache swaps. Narrow-minded employees cannot handle two tasks contemporarily because having more a sequential logic in doing things; this is sometimes also culturally related. Interruptions have a similar effect to the JOSO time in the office and have to be avoided. Suspending temporarily the task working on, due to a disturbance may lead to loosing already done VA work which has to be redone (this corresponds to a kind of rework in industry) and has to be avoided. This is especially valid for narrow-minded types of employees. In Fig. 3.15 is shown the time lost.

Interrupting a job due to a disturbance causes usually to losing the already performed VA-time, necessitating to restarting the work more or less from the beginning for a task (or for a transaction at least from the last completed task). The interruption time has therefore to take into consideration not only the disturbance time but also the consequent re-start of interrupted job. For that reason, at minimum the actual task working on should be completed. Especially the time read-up again on the interrupted job may vary from person to person and from job to job and from duration of interruption. In any case, try to avoid job interruptions.

This leads to the

Principle of "Always complete a task before interrupting the job"

in order to avoid VA loss and to avoid NVA of redoing certain steps of the work. This principle could even be enlarged to the

Principle of "If possible, strive for sub-routine or even end-to-end process completion".

In the office the man/machine separation concept of Chaku Chaku does not exist with very few exceptions of e.g. batch IT job spoolings. Being usually the employee permanent occupied on the job there is little possibility to "externalize" tasks done internally during the switch-over procedure (in SMED external tasks are tasks done during the machine is still running). That is only possible, if other persons might

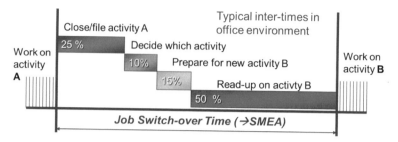

Fig. 3.16 Single minute exchange of activity (SMEA)

do the task of filing a job (concept of supervisor-assistant interaction). This needs a contingent approach according to the specific situation.

Now, is also for the office activity a switchover SMED-like technique applicable? As we have seen, an office employee has to be polyfunctional (in other words multitasking able) performing different activities such as e.g. main activities, support respectively process-required activities, report statistics, respond to contingent situations, and many more, see the Sect. 3.8. How to switch efficiently from the same type of job to another job of the same type and switching from one type of activity to another type of activity may imply different JOSO techniques because endorsing a new job or a new activity implies different preparatory work, e.g. open list, prepare sheets, i.e. generate NVA tasks. This is the reason why we distinguish both cases. We will call these office techniques "Single Minute Exchange of Job" (SMEJ) and "Single Minute Exchange of Activity (SMEA). The first we have seen described in Fig. 3.14, the SMEA is shown in Fig. 3.16, which has as an additional step consisting of deciding which activity to perform next.

If the activities are scheduled using a Heijunka activity board, then the time to decide which activity to choose can be further compressed (see Sect. 3.8). Field experience has been showing that the NVA task associated to JOSO are depending from the type of job or activity and may represent a considerable part of time. It is therefore of high interest to reduce this NVA also with the help of IT solutions.

3.7 Single File Execution and Speed

Discussing with production managers, often I heard: we need a Kanban system. Well, this is not a good idea, because a pure Kanban system usually entails supermarkets. Supermarkets increase net working capital and bind liquidity. Therefore, a pure pull-based backward chaining production from the end to the beginning of the process should be avoided because process supermarkets constitute also a sort of WIP, which should be reduced. The better idea is: we need a SPF triggered by demand-pull. A SPF is a tayloristic-based single-piece-push manufacturing mode. For the classification of manufacturing modes consult [3]. To reduce WIP the asynchronous JIT (just-in-time) should be, if possible, transformed into a synchronous JIS (just-in-sequence) [3] as it is implemented on automotive final assembly lines. The functional aim of the TPS is not establishing Kanban supermarkets but implementing SPF, although the TPS is mainly based on internal JIT make-to-stock production principle.

This leads to the

Principle of "Pull if you cannot flow"

aiming to limit WIP and to maximize response. This principle is less applicable to the office but it embodies fully the concept of Lean manufacturing.

In industry, the two of the four main manufacturing modes are B&Q and SPF. The difference for a make-to-order production principle is shown in Fig. 3.17. For a complete treatment of the topic see [3]. It shows with the help of VSM how raw material from the RMI is transformed into finished goods. In the B&Q mode the production is centrally controlled by the MRP2 or ERP system [e.g. 18, 19, 20]. In the SPF mode the production is triggered by the demand according to the TR. In the make-to-order production principle, the finished goods are not stored but ideally immediately shipped to the customer.

Generally, the make-to-stock production principle is not applicable in the transactions-based service industries (exceptions may be given by e.g. marketing brochure re-ordering), this because as already said, the products of transactions are not defined products but different parametric templates with changing (parametric) content of each product template; only the make-to-order production principle applies therefore in the office. Every order is therefore special; in industry we would talk of a customized order. In the office, I prefer to talk about a **parametric order** than a customized order, because in industry the customization is made on a reduced set of product characteristic realized in the final assembly area starting from a standard base, whereas the transactional order template is fully parametric. This means in the office world we are in the presence of a 100% batch size 1 regime. Please note, do not confound SPF with batch size 1, as personal consulting and lecturing experience has shown often happens. SPF refers to the transfer principle, i.e. a single unit is transferred form one operation to the next, instead of the entire batch; the order quantity also for a SPF may constitute of several identical pieces, i.e. the batch size. However, is a SPF realistic in an office environment? A SPF entails a tayloristic-split production process. Generally, splitting makes sense if no coordination information has to be added, but with a batch size 1 you need to apply after each job order always JOSO which then may need information for the new job. If the execution of the different operations in a process need different competencies (competence which is concentrated in different departments), the tayloristic procedural approach makes sense. However, if the coordination effort rises it makes sense to have an integrated organizational approach in place, i.e. a single person or a team within a work cell makes the whole work content. A work cell allows improved information interchange among the employees working in the cell. Figure 3.18 shows the two organizational principles, which might also be used to de-bottleneck a constraint. As a vague rule, we can state: a procedural form will apply for inter-department processes, whereas an integrated form will rather apply for intra-department routines.

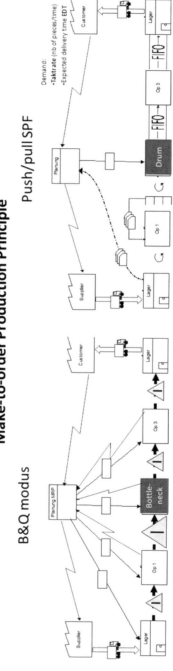

Fig. 3.17 Make-to-order production principle interpreted with B&Q versus SPF

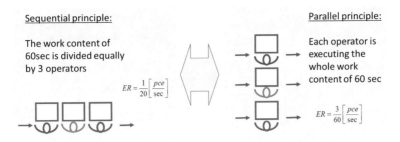

Fig. 3.18 Organizational (balancing) principles: procedural (sequential) versus integrated (parallel) [3]

We have seen that the SPF entails a tayloristic process; by the way, a SPF presents a shorter PLT than a B&Q manufacturing mode—speed is also important in an office environment. For a detailed discourse about PLT of different principles as well as related equations we refer explicitly to [3]. Due to the fact, that often in the office we have an integrated processing approach, the concept of SPF loses its meaning. We will therefore substitute the concept of SPF, which we have already briefly seen in the introduction chapter, with the concept of single piece handling (SP handling) because never obtaining a perfect SPF as defined in

$$SPF := \left\{ \lim_{n \to 1} Push\{B_k(n)\} = \lim_{n \to 1} Pull\{B_k(n)\} | CT_i = CT_{i+1} \right\}$$

Indeed, a perfect SPF can be defined as [3]

Central Limit Theorem of Manufacturing CLTM (or Definition of Perfect SPF or Manufacturing Principle Identity Theorem):

In the case of a balanced manufacturing line with equal cycle times CT at each workstation, when the transferred unit tends to one, the two manufacturing principles Push and Pull become indistinguishable, defining a perfect balanced SPF.

Corollary to the Central Limit Theorem of Manufacturing (Corollary of Improper SPF):

If the transferred quantity tends to one, but the cycle times CT are not equal (unbalanced line), this condition defines an improper SPF (or SP handling); the manufacturing principle push or pull may still be distinguished and depends from the triggering.

Instead of talking of SP handling we will introduce the concept of single file execution (SFE) which makes more sense in an office environment. SFE shows a pleonastic character: indeed, it unifies two aspects: the batch size 1 character and the single piece transfer principle. However it does not explicit if the file is processed by one single employee (integration) or tayloristic by several employees (sequentially), e.g. recommendable within an office cell but not necessarily.

In industry, to implement a SPF often the drum-buffer-rope (DBR) technique is applied, usually at the bottleneck, or at the customization point, which is Heijunka-box scheduled (Fig. 3.19). For the scheduling, we refer to the next Sect. 3.8.

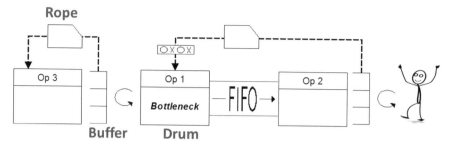

Fig. 3.19 Goldratt's Heijunka scheduled DBR technique for a SPF in an industrial environment

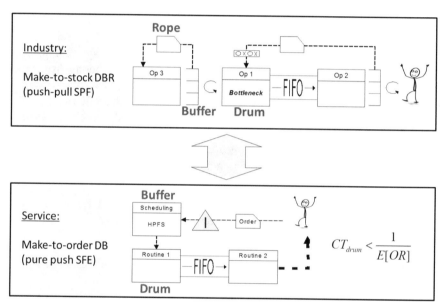

Fig. 3.20 Pure push SFE in an office environment from the beginning of a process or a routine

However, this technique of scheduling at the bottleneck or the customization point is not applicable in the office environment, or at least, is very unnaturally and leads to confusion of the upstream pulled operations. For the office, we have to apply a pure push SFE because a push/pull is really hardly implementable (Fig. 3.20). The first operation will become the drum and pushes the transactions through the organization; the balancing of teams and departments will become an essential topic and of central modeling. The backlog will be sequentialized according to HPFS (see next section).

Figure 3.20 shows the big difference of industry versus service. In industry, the buffer is replenished by the "rope" according to Heijunka-scheduled demand and pulled by the "drum". However, in an office environment the buffer is replenished by the randomly arriving client orders and is therefore "push"-governed. This "buffer" then is scheduled according to HPFS (highest priority first served) and the process

Fig. 3.21 Lean office
Conwip to stabilize WIP and
therefore PLT

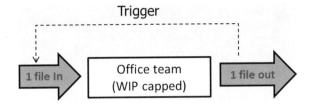

pace is imposed by the "drum". How to link routine 1 and routine 2 of Fig. 3.20 is described in the next Chap. 4. Instead of a FIFO (first in first out) SFE line, we should talk for an office environment of a FCFS (first come first served) SFE line; this is only to distinguish the type of environment, the concept is identical. The FIFO lane limits the number of orders in the lane. How the scheduling of the SFE is organized, will be treated in the next Sect. 3.8, but we can anticipate that we will focus on HPFS. The balancing of the routine or the whole process will be treated in the Chap. 4. Please note, the FCFS line has to be executed as pure SFE, i.e. as soon as a file is completed at an operation it has to be forwarded immediately to the next operation not waiting to be batch transferred together with other files; the artificial forward batching would only entail a delay. This is typical of paper-based internal post-office file distribution. Of course, in order to make sense, the number of files, as well as the requested delivery time, has to match.

This leads to the

Principle of "SFE forwarding instead of batch forwarding to reduce PLT".

To create a flow with stable PLT the WIP has to remain constant. This is called "CONWIP" (constant WIP) aka "generic pull". The generic pull principle is applicable to any manufacturing mode, considering production as a black box manufacturing system and corresponds in fact to a controlled push. The same concept is also applicable to an inter-department process (i.e. between different departments). For the case of an integrated sub-routine executed by a single employee or team, the concept applies for a single file (Fig. 3.21). In the next section, we will come back how to integrate CONWIP into the prioritization approach.

If the generic pull is applied to an inter-department end-to-end process, it stabilizes the WIP and therefore the PLT. This of course is only valid in the average due to the changing product/service mix and product/service complexity. Due to the changing content of a parametric template, the CONWIP technique is ideal to remain lean, although considering the whole process as a black box. It is necessary to define the WIP cap in order that every employee has always a WIP in front, small but necessary to buffer the different workloads at different office cells or single employees (see the concept of "office cell modus operandi" in Chap. 4). If the ER is temporarily smaller than the OR, a backlog WIP will build-up at the beginning of the process and between different office cells.

A lean transactional process can be separated into four macro phases:

1. order entry

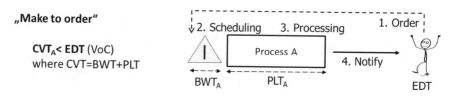

Fig. 3.22 For OTD the CVT has to be shorter than EDT

2. prioritization/scheduling at the begin
3. executing the process
4. notification to the customer (Fig. 3.22).

It is of utmost importance, that all mandatory data is available at order entry moment; generally, no incomplete order has to enter the process. To respect the OTD the customer visible time (CVT) made up of the backlog waiting time (BWT) and PLT must be shorter than the customer's expected delivery time (EDT), which corresponds to the VOC, as we have seen according to Eq. 3.3.

If the generic pull is applied to an intra-department sub-routine (i.e. within the same department), it stabilizes the WIP of the department and therefore the PLT of the sub-process. This is important because the entire process may be the sum of different sub-routines; if a sub-routine is delayed, the whole process will be delayed. In the case, the sub-routine is executed without interruption by a single employee (integrated organization), then the PLT of the sub-routine corresponds to the CT of the sub-routine (work content). In the following, we will define some equations according to the contingent situation that will appear.

Although we have a SFE, for a process spreading according to swim-lane VSM over different departments, WIP will build-up at the interface of the hand-over between the departments j and j−1 respectively between the cells. This corresponds to a capped capacitive FIFO lane in industry but with the difference that in the office the WIP, if advantageous, might be rescheduled with a Heijunka box (see next Sec. 3.8). Different from industry, this WIP is important because it will level the variability of execution of jobs with different size of content having different CT. Moreover, WIP may be necessary due to flexible working hours of employees; in this case, SFE can only be implemented in the common period of enterprise block time, i.e. blocked time slot for all employees. The whole process lead time, which corresponds to a value stream (and corresponds to the customer visible time CVT), is the sum of the sub-routine lead times (RLT) including also the BWT of order entry and can be computed according to Eq. 3.12, which is derived from Eq. 3.3b; it shows a stochastic characteristic due to non-deterministic job contents. If the RLT show an upper specification level (USL) then the VOC of OTD can be maintained every time. Backlog waiting time (BWT) is the result of differences between average order entry E[OR] and the E[ER] of CONWIP managed process. Please notice, the prioritization of orders is especially applied at the order entry point to have the full benefit, whereas in the department the main tool is the Heijunka box. WIPT is the

WIP waiting time depending also of an eventual additional prioritization scheduling at department level.

$$E[CVT] = BWT(orderentry) + \sum_j WIPT_{j//j-1} + \sum_j E[RLT_j(dept)] \quad (3.12)$$

For a sub-routine performed within a department j we have to distinguish two cases: a) a tayloristic-based office cell organization synthesized in Eq. 3.13 and Eq. 3.14 on the one side and on the other side b) an integrated single employee work organization (Eq. 3.15). For the case a) generally applies the same thoughts as for the process, but whether a WIP between the employees will build-up depends on the cell organization, i.e. if it is WIP-capped managed. If the department sub-routine transits only one cell applies Eq. 3.13 (here the case of a non-balanced cell forming WIP between the CT_i), if it transits more than one cell, the department RLT results to be as shown in Eq. 3.14.

(a) PLT of a sub-routine (RLT) performed within a cell (tayloristic)

$$E[RLT_j(cell)] \approx BWT_j(Heijunka) + \sum_i ((WIP_i + 1) \cdot E[CT_i]) \quad (3.13)$$

$$E[RLT_j(dept)] \approx BWT_j(Heijunka) + \sum_k WIPT_{k/k-1} + \sum_k \sum_i ((WIP_i + 1) \cdot E[CT_i])_k \quad (3.14)$$

(b) PLT of a sub-routine performed by a single employee (integrated)

$$E[RLT_j(employee)] = BWT_j(Heijunka) + \sum_i E[CT_i] \quad (3.15)$$

The Eqs. 3.13 and 3.15 compared to 3.14 show clearly, that the integrated single employee performing an entire job according to the "end-to-end" principle is advantageous with respect to the cell RLT in the case that the tasks among the employees of the cell are not well balanced. We see here another indication, that the integral (parallel) concept of Fig. 3.18 in an office is more appropriate, in general; this is not the case for industrial high performance manufacturing.

The equations give also evidence for how important it is to show the WIP not only in industry manufacturing processes but also in transactional swim-lane VSM presentations; drawing swim-lane VSM in the office is mandatory and stands at the base of every Lean office transformation. Usually it is the WIP, aka Muda, which makes the lion's part of PLT. Equation 3.13 shows how WIP increases PLT. Stabilizing WIP is therefore mandatory in order to have a constant PLT in average; a stable WIP allows predictable and repeatable PLT.

Stabilizing the WIP corresponds to the

Principle of "Output triggers the release of new orders or, input equals output will stabilize PLT".

Such as in industrial environment the bottleneck machine has always to be busy, if comparing the office employees to the machines, the employees have always to

Heijunka Box of Cell X/ABC										
Part	Daily production requirement	06:00	07:00	08:00	09:00	10:00	11:00	12:00	13:00	14:00
A	200	50		50		50		50		
B	100		50				50			
C	100				50				50	
additional Kanban										

Fig. 3.23 Industrial Heijunka scheduling box with pitch of 50 units

be busy with VA tasks. A minimal and capped WIP will be necessary to assure an operational work reserve for the employees; this leads to the

Principle of "Keep employees always busy with VA-activities, i.e. eliminate NVA-activities"

to reduce the incidence of Muda related activities. These concepts are fundamental to implement a lean office cell.

3.8 Prioritization Pull and Effectiveness

The main differences of a B&Q mode versus a SPF mode, apart the transfer principle such as batch or single piece, lies usually also in the triggering of production: in B&Q mode, production is scheduled by push ERP systems, in SPF mode within the TPS production is TR-controlled and JIT-pulled on demand. The TPS is a Kanban-triggered make-to-stock production with small supermarkets. These supermarkets are strategic buffers, which separate CONWIP-pulled manufacturing cells with different cycle time (CT). Kanban is how the pull is implemented. Kanban may be cards, lights or acoustical signals, but increasingly also electronic signals are used. Production starts only in the presence of a Kanban and with defined quantities associated to a Kanban. Therefore, the WIP will be limited in the system and the PLT becomes deterministic and not variable; a constant PLT is essential for JIT. The application of a pull system is conditioned to a low variability of orders represented by the coefficient of variation measure, i.e. the following condition should be verified

$$CV[OR] = \frac{SD[OR]}{E[OR]} < 1$$

In JIT industries, the scheduling of different orders is made with the help of a so called Heijunka-box respecting the concept of Mura. Mura means leveling demand, splitting different production orders into equal orders sizes assigned to deterministic time slots, called pitch. Each work cell usually has its own Heijunka box. The concept is shown in Fig. 3.23.

Fig. 3.24 JOSO and Heijunka activity scheduling

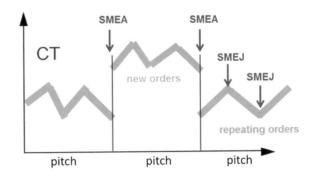

Whereas in Western B&Q modus large production orders are preferred to reduce the incidence of NVA changeover (C/O) time related to production time, the TPS prefers small and equal batches. The logic behind, is that the company has to be able to deliver all products within a short timeframe; producing a large lot would block capacity. This means that large orders are split to allow also smaller lots to be delivered; this necessitates short C/O times.

In an office environment we can use the same technique to schedule orders, but it has to be adapted. In the office, the Heijunka box is not primarily used as a leveling board, because we have not the problem to split orders of different batch sizes, but rather as an activity-scheduling box for multitasking cell/departments. Indeed, teams in an office process different single files; although the files may be of different resource absorption, they should be executed according to the "end-to-end" principle. Further, due to the multiple different tasks and sub-routines, which teams of a department have to perform, the scheduling is more oriented to alternate the different kind of tasks and sub-routines, as well as ancillary activities which have also to be performed during a day, each activity may have its own WIP. To perform JOSO (with the technique SMEA, but also SMEJ) is therefore crucial (Fig. 3.24).

The problem is therefore to allocate different time slots to different activities in order to perform all the necessary multiple works that teams of a department have to fulfill. It is useful to structure the office Heijunka board in three sections: activities performed within cell, activities performed within department, and individually performed activities. This concept is shown in Fig. 3.25.

In the previous sections, we have stressed the efficiency aspect of the process, i.e. doing the things right, now we have to look at the effectiveness, i.e. doing the right things. The problem of limited resources, i.e. employees, facing a daily workload that should be processed as quickly as possible is a constraint; hence scheduling of work is necessary. An important topic is therefore how to prioritize different orders or activities. Usually a FCFS scheduling applies. Independent of the type of activity, commercial orders or ancillary tasks, it is important to schedule jobs according to priority. Priority has a two dimensional criticality: importance and urgency (Fig. 3.26).

The important and at the same time urgent jobs get the highest priority and have to be processed first. Then will follow the urgent jobs and then the important jobs.

Heijunka Box of Dept/Cell X/Y1							Day 00						
Activity	Daily approximate requirements	08:00	09:00	10:00	11:00	12:00	13:00	14:00	15:00	16:00	17:00	18:00	
within cell													
- meeting													
- operations routine													
- support routines													
- Kaizen													
within department													
- meetings extra													
- projects													
- non structured routines													
individual													
- e-mails													
- trainings													
- reports/statistics													
additional Kanban													

Fig. 3.25   Office Heijunka scheduling board for activities

Fig. 3.26   Office activity prioritization with the decision board according to HPFS

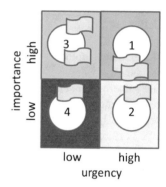

This leads to the HPFS (highest-priority-first-served) principle. For all these different jobs, the JOSO principle apply. However, how to apply these office boards? Firstly, before starting in the morning, the team has to get an overview of jobs to be performed during the day, secondly the jobs are prioritized with the help of the matrix decision board, and thirdly they are scheduled with the help of the Heijunka activity scheduling board. Further information how a team will interact will be discussed in Chap. 4 and Sect. 5.1 of the Chap. 5.

The presence of limited resources in concomitance of concurrent jobs to be performed leads to the

Principle of "Install priority-pull or HPFS scheduling in order to increase effectiveness".

With this matrix we put the emphasis not on "doing the things right" but "doing the right things". This corresponds to the

Principle of "Put-efficacy-over-efficiency"

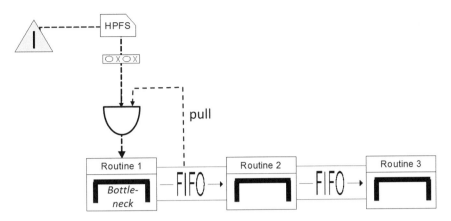

Fig. 3.27 Single point of prioritization at the begin triggered by the "input equals output" principle

or HPFS Principle (highest priority first served) leading finally to the overall effectiveness of the organization. "Doing the things right" is assured by normalized work, Poka Yoke, and TEM.

In order to do "the right things" also at the right time, a job release rule has to be established. The order backlog is prioritized daily, and then, together with other activities, scheduled on the Heijunka activity board. The job release will comply with CONWIP. As soon as a file exits the first cell, the job is pushed as SFE through the next departments (cells). At the same time a new job is released from the scheduling board (symbolized with the "and"-gate in Fig. 3.27). To maintain the WIP constant, it is the bottleneck operation, which triggers the release of the new job or activity. If a job needs to be suspended, e.g. due to missing data, the cell potentially cannot continue to work. Therefore, additional rules have to be established, or a larger WIP cap is necessary, or better, how it should be, no file with incomplete information might be allowed to enter the process. This should be assured at the order entry instance, before being prioritized. How to conceive a cell will be treated in the next Chap. 4.

It has to be mentioned that at any transaction handover between two cells (or departments) for multitasking cells in concomitance of a Heijunka scheduling box also a prioritization re-scheduling might be applied (Fig. 3.28). This might be useful in contingent situation, but usually the prioritization is only done at begin of the value stream. Indeed, in cell 2 of Fig. 3.28 in addition to sub-routine A two further sub-routines C and D are executed and must be scheduled. However, in cell 3 no new sub-routine is entering and therefore a new prioritization is not necessary; they queue according to FCFS principle.

It is important to mention, that in those cases where services are stratified in categories such as e.g. in express orders, priority orders, or normal orders with different EDT, no re-prioritization should be applied and the prioritization should be defined

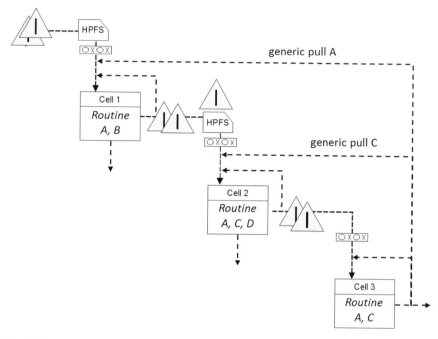

Fig. 3.28 Multi-point of HPFS re-prioritization of WIP in the case of multi-routine office cells

only once at the beginning. The WIP-cap along the process will stabilize the routine lead times to assure final customer OTD.

Due to the difference between a manufacturing and office work place but also due to the intrinsic nature of demand characteristic and how demand originates, in most cases, the supermarket Kanban principle cannot be transferred as is to the office world. The pull-trigger to work might be rather the "boss-pull", or more usually the "customer-pull"; however afterwards, generally, the virtual product, i.e. the transaction, will be pushed through the process on a FCFS flow basis! The processing system can be assimilated to a generic pull system (which is a controlled push). Please note, the sometimes installed order board, named Kanban task board, situated in-between departments in office application is not a real pull system, as often pretended. A real and pure pull mode such as "single piece pull" for unbalanced lines [3] is triggered from the line-end towards upstream operations; the downstream withdraws from supermarket are immediately replenished by upstream. The final, or customization, call-offs select contingently among a limited product-mix stored in a supermarket. This is not the case in the office because no intermediate replenishment-pull supermarkets can be conceived, due to the uniqueness of each single order file. Remember, a supermarket is a strategic buffer with a limited and standardized product/component-mix allowing to operating JIT. In office, the triggering is clearly at the beginning of the line pushing the product/service through the organization (make-to-order production principle starting "ex novo"). However, in industry the triggering to produce a certain product or component is made by a product-defined downstream production Kanban at the end of the line. In the office the real triggering what to produce is not indicated by the "exits" to provide a certain service/product (the "input equals exits" guarantees only the WIP-cap by CONWIP technique) but which product/service to produce is dictated through the HPFS principle at the begin of the line! The selected product/service is then pushed through the departments based on FCFS flow (eventually rescheduled by intermediate HPFS boards if the downstream department performs also other routines). This is another indication that the TPS Lean tools are still valid for service and administration environments but they have to be reinterpreted and adapted.

References and Selected Readings

1. Inside Paradeplatz: UBS-Industrialisierung: Tod oder lebendig? News of 7 Dec 2014, inside-it.ch
2. Rother, S.: Learning to See. LEI/Cambridge Center (2003)
3. Rüttimann, B.G.: Lean Compendium—Introduction to Modern Manufacturing Theory. Springer (2017)
4. Chen, C.: Value stream management for lean office—a case study. Am. J. Ind. Bus. Manag. (2012)
5. Bonaccorsi, A., Carmignani, G., Zammori, F.: Service Value Stream Mapping (SVSM): developing lean thinking in the service industry. J. Serv. Sci. Manag. (2011)
6. Goldratt, E.: Theory of Constraints. North River Press (1999)
7. Visco, D.: 5S Made Easy: A Step-by-Step Guide to Implementing and Sustaining Your 5S Program. CRC Press (2016)
8. Fabrizio, T., Tapping, D.: 5S for the Office: Organizing the Workplace to Eliminate Waste (2006)
9. Locher, D.: Lean Office and Service Simplified: The Definitive How-to Guide. CRC Press (2011)
10. Inspire AG: Lean Six Sigma Black Belt curriculum. Inspire academy (2014)
11. Martin, T.D., Bell, J.T., Martin, S.A.: The Standardized Work Field Guide. CRC Press (2017)
12. Pyzdek, T.: The Six Sigma Handbook. McGraw-Hill, New York (2003)
13. Töpfer, A.: Six Sigma—Konzeption und Erfolgsbeispiele für praktizierte Null-Fehler-Qualität. Springer (2007)
14. Nakajima, S.: Introduction to TPM: Total Productive Maintenance. Productivity Press (1988)
15. Shingo, S.: Zero Quality Control: Source Inspection and the Poka-Yoke System. Productivity Press (2018)
16. George, M.: Lean Six Sigma: Combining Six Sigma Quality with Lean Speed. McGraw-Hill (2002)
17. Shingo, S.: Quick Changeover for Operators: The SMED System. Productivity Press (1996)
18. Lunn, N.: MRP Integrating Material Requirements Planning and Modern Business. McGraw-Hill (1992)
19. Petroff, J.N.: Handbook of MRP 2 and JIT: Strategies for Total Manufacturing Control. Prentice Hall (1993)
20. Adam, M., Schäffler, S., Braun, A.: Lean ERP—How Lean Management tools are supported by ERP-systems—an overview. In: Piazolo, F., Felderer, M. (eds.) Multidimensional Views of Enterprise Information Systems. Springer (2014)

Chapter 4
Office Cell Design

In Chap. 3 we have learned how to interpret the basic Lean tools for the office environment. The accent has been put on the interpretation and not on the application. Knowledgeable Lean experts are necessary to implement this tool system to the contingent office situation taking also into account the present office culture. In this chapter, we will tackle the approach to implement a lean office cell by conceiving the practical set-up of the theoretic concepts seen in Chap. 3 building a lean office cell. With that, the first step of office industrialization becomes reality.

4.1 The JIT Governing Logic

What means just-in-time (JIT)? How can it be applied to the office? Different definitions of JIT exist according to [1]. Indeed, according to Robert Hall JIT means stockless production and zero inventories. However, already in 1983 Edwards (according to [1]) describes JIT with the seven zeros (zero defects, zero lot size, zero setups, zero breakdowns, zero handling, zero lead time, zero surging [1]). For a more scientific definition of JIT we refer to the following theorem [2].

Cardinal Theorem of Lean (or JIT Theorem):

Necessary but not sufficient requirement for a JIT production, i.e. the time dimension to implement the Lean vision of the right product, at the right place, at the right time, is to strive for batch size one, intended as transfer unit, i.e. a SPF, to minimize WIP and aiming to have a balanced line.

Having a SPF means minimizing WIP, minimizing WIP means to reduce PLT and reducing PLT means potentially to have a JIT manufacturing regime implemented. JIT can also be translated to a more comprehensive exhaustive mathematically formulated model expressed as follows, adding the condition of the Theorem of Throughput

$$JIT := \left\{ pull(SPF) = \lim_{n \to 1} B_k(n) \,\middle|\, minPLT = \min_{CT_i = CT_{i+1}} WIP, \sup CT_i = CT_b \le TT \right\}$$

© Springer Nature Switzerland AG 2019
B. G. Rüttimann, *Transactional Lean: Preparing for the Digitalization Era*,
https://doi.org/10.1007/978-3-030-22860-6_4

where pull of a single piece flow (SPF) means that the transfer unit has to tend to one with the conditions to be observed are: that to minimize PLT one has to limit WIP and that the CT at the bottleneck has to be faster than the TT. Please note, JIT is not a synonym of OTD. OTD covers the specification aspect of a customer request to have in time deliveries whereas **JIT covers the theoretic dimension of Lean being a customer-pull manufacturing theory**. The OTD equations have been enunciated in Chap. 3. It is important to notice, that TR usually has a reduced variability, especially in automotive industry; we can therefore state that $SD[TR] = 0$. In other industries, not a constant TR but a variable order rate OR is observable, i.e. $SD[OR] > 0$. For other industries than automotive, e.g. such as service industries, it is better to talk about an order rate OR than a takt rate TR. In Fig. 4.1 is shown the relationship among the driving variables of Lean.

Figure 4.1 shows clearly that the TPS is centered on TR and the observation of the expected delivery time (EDT). It shows also the metric to implement the concept of cellular manufacturing. We see again the key concept of CT at the bottleneck (CT_b) as well as set-up time (ST) in the case of a mixed product cell. The other important variables are cell turnover time (CTT), product interval time (PIT), and the batch-size of product k (B_k). For further information, we refer to [2].

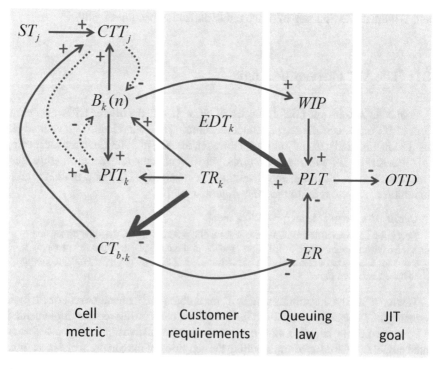

Fig. 4.1 The central importance of TR among the driving variables [2]

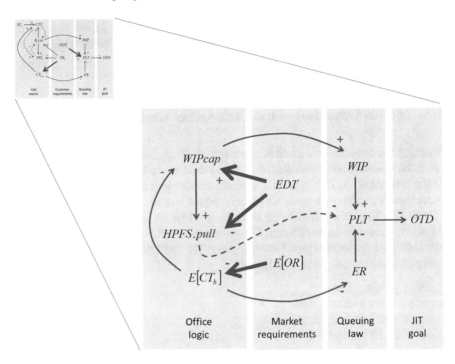

Fig. 4.2 The industry-derived governing JIT logic of Lean Office

Nevertheless, there is no difference conceptually to implement a Lean office cell, only the contingent situation has to be observed. Indeed, due to the prevailing push principle in the office, to observe OTD with the EDT, the HPFS has to take the PLT in consideration. Therefore, if we consider the JIT governing principle of Lean, we can refer to Fig. 4.2 for the office application.

It shows the central importance of market-imposed OR and EDT for each order. The respect of order rate (capacity dimension) as well as delivery time (speed dimension) is also in office application essential. The observation of the queuing law is a consequence of the applied principles of the office cell, as Fig. 4.2 clearly shows. To observe customer's OTD, PLT and ER have to be appropriate, as synthesized in Chap. 3 with Eq. 3.3. To respect expected delivery time (EDT) and to match OTD not only the capacity of the office cell but of the whole process have to be large enough; in addition, the WIP has to be capped, i.e. limited, and the entering orders have to queue according to highest priority first served principle (HPFS). Please notice, the dashed line to PLT is only symbolic because not shortening PLT; PLT is given by WIP and ER. In reality, the prioritization avoids queuing time and the order is processed with priority. Matching OTD is mandatory! This is the governing framework of a lean office cell.

In the next two sections, we will cover the practical set-up of Lean in an office environment, meaning a new industry-derived way to work, yielding a higher performance.

4.2 Modus Operandi of the Lean Office Cell

An industrial manufacturing cell is a self-contained organized entity having all necessary equipment and blue-collar workers to produce in a SPF a determined scope of defined components. Different cells are linked with supermarkets to form a complex manufacturing system producing a product, which can be shipped; for further details consult [2]. The reason for having different interlinked manufacturing cells is that the work cells may have different CT or better cell PLT and cannot be integrated into a single transfer line paced at customer's TR. An industrial manufacturing cell is implemented in five steps as follows:

- define product family
- understand operator's work
- layout equipment for flow
- staff the cell to meet TT
- implement pull system.

For further details and correct staffing, we refer to [2]. Now, is it possible to transfer this concept to an office department? If we want to increase productivity of an office team, it is mandatory to introduce industrial organizations techniques. The question is not only how to do it but why it has not yet been done. Indeed, why should shopfloor employees comply to strictly time paced manufacturing procedures and office employees work at its own discretional pace, implying Muda. The answer lies not only in the pretended different kind of work to be performed and from variability of content. The reason rather lies in the laissez-faire regime and ignorance of managers having never approached the problem in an industrial way relying only on consultants' restructuring techniques when the situation becomes critical. We have to pay attention, usually the highest cost items in a P&L statement are materials and personnel cost; in service industries, no material is listed in the P&L but even more personnel cost—the need to act is urgent, and the focus of optimization techniques have to be transferred from shopfloor to the office. The crucial question is not how many employees to lay-off; the question is how to increase productivity to offer a better service and grow business. Although it will not be possible to introduce standardized work as we have already mentioned in the Sect. 3.3 it is possible to boost productivity by a stricter work organization. However, what is an office cell? We can define it as follows:

Definition of an office cell:

An office cell is a self-contained (staffed with employees having the required competence and with the necessary number of FTE to meet OR) and self-controlled (working according to a leveled HPFS-principle) mixed-service unit performing a sub-routine or an end-to-end process according to a defined modus operandi by applying a SFE forwarding principle.

Prior to install an office cell, it is advisable to be clear about the activities, routines and tasks, which are performed in the department. This is not only necessary to select the routines to be transferred to the office cell, but to have an overview of all activities to be included into the Heijunka pitch. For the sub-routines, and the processes, it is advisable to draw upfront the swimlane-VSM in order to have an overview of the main activities to be processed in cells.

When makes it sense to implement an office cell? Above all, we have to be clear: it would be an illusion to believe to implement a perfect SPF in an office. The changing parametric content of each order leads to variable CT at each workplace so that we have to talk of an E[CT], i.e. an average CT of most of the tasks. To limit variability SD[CT] it is advisable to define classes of complexity within a product/service. The classes will not only reflect the difficulty (which defines the necessary competence of the employee) but also reflect the necessary CT (which defines the resource absorption). Different from manufacturing, where a SPF without FIFO lane between workstations at the limit has no WIP and therefore PLT corresponds to the work content, this is hardly achievable in an office cell. On the contrary, if the job is performed by a single employee (integrated organization) then as we have seen in Sect. 3.7 the sub-routine lead time (RLT) equals the work content of the job because the employee will do task after task without stopping (begin-to-end job). If the job is performed by a team, ideally within a cell (tayloristic organization), then as we have seen in Sect. 3.7 the RLT may comprise WIP as shown in Eq. 3.13. Indeed, the CT of the different tasks within an office cell will be difficult to balance.

There are different ways to implement and run an office cell, what we will call the modus operandi of the office cell. The modus operandi is made-up of different implementation aspects such as

(a) scope (intra- or inter department)
(b) organization (CT-balanced or WIP-capped)
(c) interaction (procedural or relational)
(d) vehicle (paper-based or electronic file-based).

The last aspect allows to implementing also a virtual office cell. Aspects regarding product type, layout, balancing etc. are treated in the next Sect. 4.3. Let us now look more closely at each dimension of the modus operandi.

(a) Scope of the Lean Office Cell

In Chap. 3 we have defined a sub-routine to be a part of a process usually confined within a department whereas a process goes through different departments typically an end-to-end transaction. We can therefore describe the scope of a Lean Office cell as follows

– Intra-department cell for sub-routines ("within" department)
– Inter-department cell for end-to-end processes ("between" department).

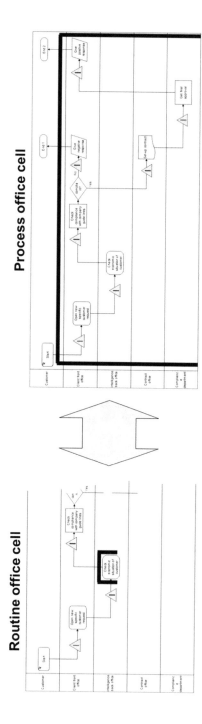

Fig. 4.3 Scope of a Lean Office cell: department sub-routine versus end-to-end process

In Fig. 4.3 the concept is shown with the help of a swimlane-VSM.

It shows that within a process not all departments have to be cell-organized, this will be the case especially at the beginning of the Lean Office deployment. However, the representation of an end-to-end process performed within an inter-department cell is difficult to represent with a traditional swim-lane. This shows further that for entire end-to-end processes performed within an office cell, swim-lane optimization technique fails. A single cellular manufacturing unit is not any more representable by VSM. Here concepts of industrial standard work apply, of course adapted to the need of an office (see Sect. 3.3).

The intra-department cell is by far less difficult to implement because the concerned employees are already working together. To make work together different employees of different departments is much more difficult. This is not linked to the employees itself but due to the need to break with traditional scheme. A new office has to be installed where grouping together different employees from different departments in order to work together in a different way. However, this different way will speed-up PLT dramatically. This leads to the

Principle of "Inter-department office cells will speed-up PLT of end-to-end transactions".

It has to be mentioned, that the inter-department scope of an office cell may solve the problem of interface issues between different departments. Indeed, long PLT as well as quality issues and "rework" originate often due to badly defined interfaces [3].

(b)   Organization of the Lean Office cell

Two possible organizational modi are imaginable how to run the cell:

– the CT-balanced office cell with a paced SFE,
– the WIP-capped office cell with a non-paced SFE.

In Fig. 4.4 is shown a schematic representation of the concept. In the CT-balanced cell organization every employee is working on his own file which is passed to the next employee according to the determined pace. It is obvious, that the work content has to be comparable to show similar CT having low variability; this will be difficult in reality. The files are transferred in a push mode. If there is too much unbalance, the time can be used to perform very short and simple tasks, such as reading e-mails. In the WIP-capped cell organization, the number of files is larger than the number of employees to load every employee avoiding inactivity. The files are FCFS pushed to the next employee. Due to the CONWIP technique realized with the "input equals output" principle the lead time will have an E[PLT] and will rather be constant.

Generally, whether WIP will build-up in a cell depends if a CT-balanced office cell (with paced SFE) or a WIP-capped office cell is envisaged. In the CT-balanced office cell the number of jobs is equal to the number of employees; the inactivity of some work places due to non-balanced CT can be avoided by JOSO to other temporarily limited activities. If a WIP-capped lean office cell is implemented then the number of jobs is larger than the number of employees. A certain level of WIP will be

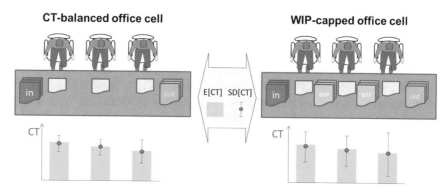

Fig. 4.4 Schematic comparison of the two cell organization modi

necessary, although capped, to adjust unbalanced contents and tasks of the employees with consequent longer PLT. However, as simulations have shown, it is advisable in presence of cycle time variability to adopt a WIP-capped cell model to maximize exit rate as simulations have been showing [2]. Indeed, WIP will decouple the single operations and allow each operations to work at its maximum ER. A CT-balanced cell will be hardly implemented because too difficult to have equal work contents in all jobs. A virtual cell will require in any case a WIP-capped organization whereas a physical office cell could also be implemented with a CT-balanced organization. It is perceivable, that the advantage of an industrial manufacturing cell is not fully replicated in the office, because the integrated organization of doing the job by a single employee is the most performant expressed in PLT, provided the employee has the competence. The capacity adjustment is done with additional employees working in parallel which concept is not possible in industry because this would entail investments in additional identical equipment. Therefore, in industry the capacity adjustment is done by splitting the bottleneck in sequential operations, whenever possible.

This leads in an office environment to the

Principle of "Advantage of integrated job execution over tayloristic-split job execution".

This is quite different from industrial manufacturing cells where a tayloristic splitting is increasing productivity by specialization. However, in the case of large department sub-routines or inter-department processes, i.e. multistage processes where different competences are necessary, it makes sense to install a tayloristic-split office cell and implement the SFE with minimal waste. The need for coordination is reduced and the exchange of information is facilitated in a cell.

(c) Interaction within the Lean Office cell

Another aspect is how to interact with other employees involved in performing the service or executing a job. Generally, two possible ways exist:

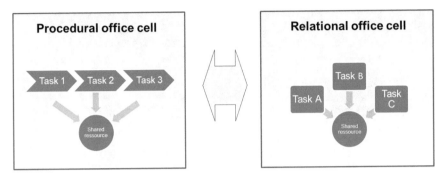

Fig. 4.5 Schematic comparison of procedural versus relational interaction

– Procedural cell
– Relational cell.

In Fig. 4.5 is shown a schematic comparison between the two possible ways. The procedural way is the common logic sequential way to work which is also used in manufacturing industry, often called workflow-type. Relational cell interaction makes sense when parallel work is possible.

Whereas in industry only procedural cell exist, in the office environment also relational cell might be appropriate. It has to be mentioned, although of these two types of interaction have never been explicitly talked in offices, they have been co-existing. These two types of interaction are also given from the activity to be performed, if parallelization is possible, apply relational type of cell. These leads to the

Principle of "Maximize parallelization of work as much as possible".

The advantage to work with relational interaction within a cell is, that it allows to exchanging information and rapidly update on arising issues. Important within a relational, but also a procedural cell is to have access at consistent and not redundant information source. The integrity of the database is mandatory.

(d) Vehicle used within the Lean Office cell

Although the paper-less office has been postulated already 30 years ago, we are still overwhelmed with paper and far away from this ideal situation. Nevertheless, electronic-based transaction handling is expanding. The way to work, not necessarily within an office cell but also generally, is different for each case. We can therefore define

– Paper-based file processing
– Electronic-file based office cell.

In Fig. 4.6 is shown a schematic comparison between paper-based processing and electronic file-based access to a centrally shared database. A paper-based cell implies

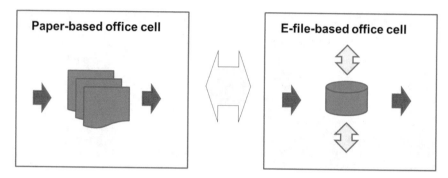

Fig. 4.6 Schematic comparison of paper-based versus electronic file-based vehicle

a procedural cell whereas an electronic file-based cell allows procedural or relational interaction.

Please note, also mixed situation may be present but these situation increase difficulty because non-consistent documents may result. Paper-based files have the advantage to be immediately physically visible whereas the electronic files are invisible and have to be made visible. Showing the WIP is important; this is directly linked to the Mieruka culture of TPS. It is immediately perceivable that an electronic file-based workflow needs the involvement of the IT department implementing a device to make WIP visible. However, the electronic-file based vehicle solution allows potentially to installing a virtual office cell, where the employees are not physically present in the same room. On the other hand, the quick interaction is not possible in this case, if not realized with permanently installed teleconferencing technology.

4.3 Six Step Approach to Implement a Lean Office Cell

How to implement a lean office cell? For that, we can partially refer to the industrial approach, but we need to adapt by adding the office cell modus operandi. We can follow the hereafter-listed six-step approach and let us have then a closer look to each single step to be worked out in order to implement the cell:

1. Define the product/service
2. Split the routine in logic tasks
3. Select the office cell modus operandi
4. Layout the office cell workplaces
5. Staff and balance the workplaces to meet E[OR]
6. Implement prioritization pull.

(1) Define the product/service: The suitable products/services for being processed within a cell have to be complex and needing different competencies, as well as consisting of at least a two stages process, otherwise a single employee can perform

the work alone at his place. Select recurring and standard-type of product/services to have a critical mass, one-offs are not suited at the beginning. It is advisable to stratify the product/service in three levels, if possible, e.g. in small, medium and large work content; simple, medium or complex WC in order to estimate the necessary CT for each category. This is necessary to be able to define standard work content units. After, one has to respond to the question: should the cell be able to perform different products/services (mixed model cell, i.e. a multi service cell) or only one (dedicated service cell)? In the first case not only SMEJ but also SMEA will apply.

(2) Split the routine in logic tasks: The splitting is necessary because structuring the process gives insights and allows to combining tasks into a workplace occupied by an office employee, eliminating Muda. Splitting the routine into tasks allows also a better handling of the interruption in the case of a disturbance factor. The time of tasks, i.e. the CT, have to be measured, defined, maintained, and improved as well as standard work content units have to be attributed to each category; this is necessary to facilitate to measure a standard ER (see Sect. 3.3 of Chap. 3). By specifying USL in concomitance with E[CT] it is possible to calculate a Cpk and to introduce Six Sigma quality management techniques also in an office to monitor and improve performance. This is also valid for routines not performed within an office cell.

(3) Select the office cell modus operandi: The selection has to be done for the four dimensions (a) scope, (b) organization, (c) interaction, (d) vehicle.

(3a) Scope: begin to explore implementation on a department cell with a sub-routine. This is easier to install and take the lessons learned to extend the concept to an entire process of inter-department scope. You need a strong support to do this because the agreement of all department supervisors involved in the end-to-end process is necessary.

(3b) Organization: in the CT-balanced cell organization (3b1), the number of jobs correspond to the number of employees in the cell. The SFE is here perfectly paced-applied. In the case of non-perfectly balanced CT of the tasks, this will lead to inactivity of employees with shorter CT. If the unbalanced time is very high compared to the previous operation (this can be verified through an informal question addressed to the pertinent employee how much time he needs to finish) it might exist the possibility by JOSO to switch during waiting to another activity; in any case, it has to be avoided to slow down the process. In the WIP-capped cell organization modus (3b2) the number of jobs in the cell is larger than the employees and a certain WIP will build-up between unbalanced employees; in this cell it is important to limit WIP and to stabilize it with the CONWIP technique in order to limit RLT. The employees will perform always the same task within the pitch and switch with SMEA at pitch-end. Of course, this is not a perfect lean cell. Both organization modi are controlled by the "input equals output" principle. In any case, modus 3b1 is generally preferable over modus 3b2 due to shorter PLT; indeed, according to modern production theory [2].

Main Theorem of Production Time (or SPF Dominance Theorem):

Regardless of the characteristic of a manufacturing system, i.e. with balanced or with unbalanced cycle times CT, the manufacturing lead time MLT is always shorter for a single piece flow SPF transfer principle than for a batch and queue B&Q transfer principle.

Corollary to the Main Theorem of Production Time (Corollary of Lead Time Limit):

In the case of a CT balanced line, the SPF principle presents the shortest achievable MLT and therefore also the shortest achievable PLT.

Lemma to the Main Theorem of Production Time (Lemma of SPF Regime):

It is always recommendable to implement a SPF principle also for unbalanced production lines; although some equipment may be waiting in a SPF, there is no lost capacity, because the bottleneck is always fully loaded and the ER remains the same but the MLT is shortened.

It goes without saying, these manufacturing laws are also valid in an office environment for a SFE. The "Lemma of SPF Regime" tells us explicitly to strive for SFE although for non-balanced CT; however, different from the machines, the employees have always to be occupied, due to the intrinsic difference of employees compared to machines!

(3c) Interaction: the interaction of employees in an office cell is more frequent than in manufacturing due to the less deterministic work content. And exactly this information exchange is advantaged in an office cell, especially for inter-department end-to-end processes. A raised question can be answered informally in a very short time (pay attention to not disturb other employees in an inappropriate instant needing to interrupt their task working on; rules have to be established). Whether the interaction is procedural, i.e. sequential, or relational, i.e. parallel (parallel is perhaps the inappropriate word in this context, better is independent not necessarily sequential), depends from two factors: the existence of a deterministic prescribed workflow and the possibility to advance on the process by several employees at the same time. The procedural interaction is easier to implement following a logic of sequential accomplishment. However, parallelization speeds-up PLT and therefore it is to prefer whenever possible. Petri net modeling is specially suited to control parallel processing and to control the advancement of the process (Fig. 4.7).

A Petri net is made-up of four elements: status, also called positions P, transitions t, directed arcs, and tokens. A process starts only in presence of a token on a position (e.g. P1 of Fig. 4.6). If a transition shows a bifurcation (e.g. t1 of Fig. 4.6), the token is doubled and allows parallel processing. The parallel transitions t2 and t3 are independent and advance according their own rules. Transition t4 will only be triggered if both positions, i.e. P4 and P5, are provided with their token, i.e. the transitions t2 and t3 are completed. For further information, consult e.g. [4]. This process representation is especially useful for electronic-based file processing.

(3d) Vehicle: In many offices the predominant vehicle is still paper, but electronic files are rapidly expanding. The change from traditional work organization to cellular office organization might give the opportunity to change also the support vehicle. However, changing two aspects, organization and vehicle, might complicate the Lean Office transformation. It is perhaps more appropriate to keep the paper-based vehicle while implementing the office cell, and once the cell is well running to change to an electronic-based file handling. This reflects the continuous improvement culture also in an office environment to make small but steady improvements.

(4) Layout the office cell workplaces: Presently, the layout of office workplaces do not follow uniform principles and even less workflow-orientation but the desks are

rather arranged according to functional organization-affinity. Alternatively to the "personally-inspired" desk layout can be seen the well-structured cellular layout. However, the layout of the desks depends from the chosen organization and vehicle modus. Figure 4.8 show the layout according to the contingent situation.

Please note, a procedural interaction is not forcedly necessary and the layout concept shown in Fig. 4.8 is, at the limit, also valid for a relational interaction. Figure 4.8 shows clearly that in the case of a paper-based and CT-balanced work organization cellular layout is mandatory. This is the most similar case of working nearly like in an industrial manufacturing cell. For the other two yellow-colored cases,

Fig. 4.7 Petri net to model
concurrent processes

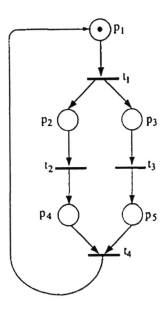

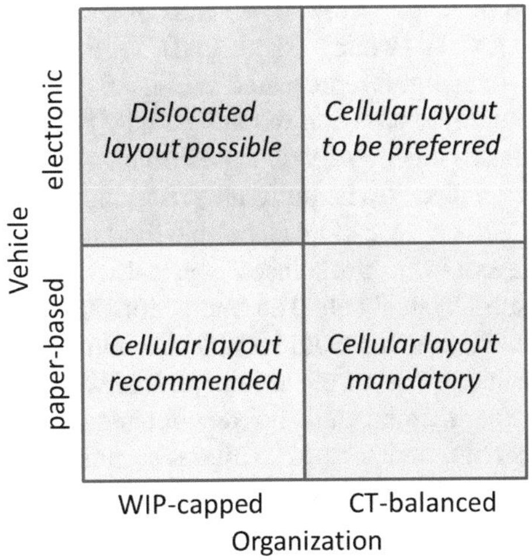

Fig. 4.8 Office layout
according to office cell
modus operandi
(organization/vehicle)

	WIP-capped	CT-balanced
electronic	*Dislocated layout possible*	*Cellular layout to be preferred*
paper-based	*Cellular layout recommended*	*Cellular layout mandatory*

Vehicle

Organization

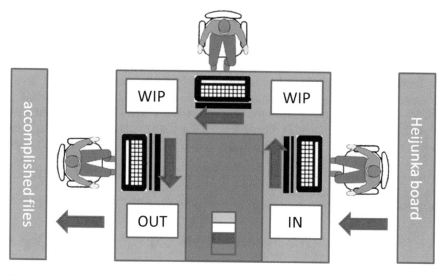

Fig. 4.9 Procedural paper-based office cell with visible WIP

cellular layout is recommended because interaction for information exchange or
informal up-dates is facilitated. Figure 4.9 shows a paper-based procedural interaction
cell. The organization of the office-cell, whether CT-balanced or WIP-capped, is not
relevant; for paper-based vehicle implement always a cell.

The Heijunka board should be placed where everybody can see it; colors may facilitate the activity identification. The storage of accomplished files should also follow
Mieruka rules (see Sect. 3.2 of Chap. 3). The layout should always be conceived to
place the desks as near as possible, which facilitates communication and paper-file
hand-over and avoids Muda of file transportation). In a paper-based transaction, the
WIP is visible; in an electronic-based file transaction the WIP is not visible and has
to be made visible. The status of a cell such as active (green), inactive i.e. individual tasks (white), issues (red), set-up or Kaizen (blue) can be made visible by an
Andon light (represented in Fig. 4.9 by the stacked colored lights). For multi-tasking
employees the cell workplace may be destined only for the specific cell tasks (this
makes it necessary of an additional workplace) or it may be a multi-functional office
workplace (to be preferred) reducing cost.

In the case of an electronic file-based and WIP-capped organization, a dislocated
layout with workplaces physically separated is possible; this would be a virtual,
non-physical cell. The interaction should be facilitated by teleconferencing equipment, and that would enable the solution even across long distances. The dislocated
solution for an electronic vehicle with a CT-balanced organization could also be
experimented but is not recommended. Although in an "electronic" cell the vicinity
is not mandatory, it facilitates communication for the handover (vocal Kanban). In
Fig. 4.10 we see the layout of an electronic-file based physical office cell. The left one
has procedural interaction, the right one has relational interaction. Central to both
is the common data base. Physically no difference is visible regarding the layout.

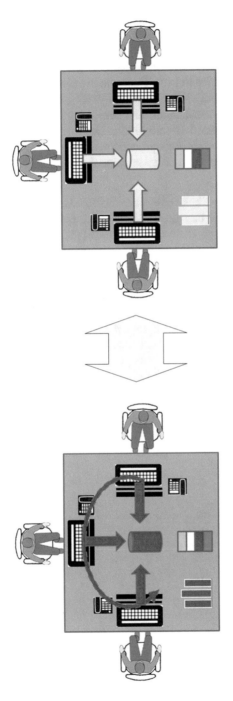

Fig. 4.10 Electronic-file-based cell: procedural versus relational

The functioning and status of the cell is shown with the Andon light; the progress of the file work-advancement in both cells has to be indicated by contingent visual management techniques.

Also in a cellular layout, the desks should be as near as possible; this is only to facilitate communication. It is also recommended to put the workplaces as shown in Figs. 4.9 and 4.10, i.e. face to face. There are examples gathering the employees at least in the same room but seating along the walls and with a conference table at the middle, which allows to exchanging and discussing information. This might also be okay for relational interaction with predominant separated work content but is less efficient (Muda). Generally, for a procedural electronic cell the vicinity of desks is highly recommended, however, for a relational electronic cell the dislocation is absolute feasible.

(5) Staff and balance the workplaces to meet E[OR]: The balancing of an office cell has not to be done only on task level but also on employee level; a time/operator balance chart of the cell will show the bottleneck (see Sects. 3.1 as well as 3.7 of Chap. 3). For that it is important to segregate a routine in self-contained tasks, which might be then classified as VA or non-VA time (Fig. 4.11). The single cycle times (CT) have to be shorter than the takt time (TT) or generally the order time.

A time/operator balance (TOB) chart should also be made on department level (which includes also individual non-cell activities); indeed, employees are multi-tasking and performing also other work than the main operational tasks. In presence of a clear bottleneck which may also be a constraint to meet E[OR] debottlenecking is mandatory. This can be done in two ways: (a) if the bottleneck is evident and it

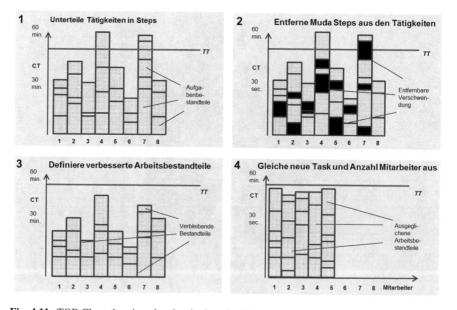

Fig. 4.11 TOB Chart showing also the single tasks [5]

is at the same time also a constraint, then a second employee has to be put at the bottleneck; (b) if the bottleneck is only slightly longer as the second longest task sequence of another employee, it is advisable first to eliminate NVA and then to balance the workload within the cell. Please note, it must result that

$$CT_b < \frac{1}{E[OR]}$$

to have no constraint. The number of employees to staff the cell (in full time equivalents FTE) can be calculated according to Eq. 4.1, where the average work content WC is divided by the expected takt time of the order rate.

$$FTE = \frac{E[WC]}{1/E[OR]} \tag{4.1}$$

The work content WC is the sum of the cycle times CT composing the routine or process (Eq. 4.2).

$$WC = \sum_i CT_i \tag{4.2}$$

Perform a Kaizen if a fraction of FTE is resulting from applying Eq. 4.1; the fraction has always to be rounded to the lower level by searching for hidden Muda eliminating NVA tasks.

(6) Implement prioritization pull: Because the work backlog should be worked-off at the end of the day, we have to work on all activities. We will use the Heijunka activity board to schedule the work sequence, which originally is also a tool to level the production batches. We will use it to coordinate the activities performed within a cell with the remaining individual activities. Before using the Heijunka board, the jobs have to be prioritized according to HPFS; this is perhaps the most important task reflecting the effectiveness of input transformation and can be considered VA time. Indeed, serving customer fast is part of the SPQR axioms. The prioritization should be done at the very beginning of order entry; an intermediate re-prioritization may become necessary in presence of multi-activity cells (see also Sect. 3.8 of Chap. 3). It is important to mention, that this prioritization and scheduling technique can also be applied for a non-cell organized department, see also Sects. 3.8 as well as 3.6 of Chap. 3.

Also for non-cell organized offices, the matrix prioritization board is decisive and should be used. For the classification in urgent/important, urgent, important, non-urgent/non-important we need also to see the frequency of the different activities in order to adapt the timeslots (pitch) to the different type of prioritization (see Sect. 3.8 of Chap. 3). On the other hand, the Heijunka way to work needs a run-in phase to reflect the correct allocation to real time needs. The decision matrix should only contain customer-related jobs, the Heijunka prioritization board should contain all activities, i.e. also recurring activities.

An additional element to take into consideration is the multi-functionality of employees. Such as the general superiority of mixed product cells over dedicated mono product cells, in general, polyfunctional (multitasking) employees present an advantage. Indeed, in the case of illness or other absenteeism or less workload, according to the mean square error of the Central Limit Theorem of statistics, the polyfunctional employees show a lower variability in occupation being better loaded.

Now, how to run the office cell we will see in Sect. 5.1.

References and Selected Readings

1. Hopp, W., Spearman, M.: Factory Physics, International Edition 2000. McGraw-Hill
2. Rüttimann, B.G.: Lean Compendium—Introduction to Modern Manufacturing Theory. Springer (2017)
3. Adam, M.: Schnittstellenmanagement—Vom Suboptimum zu einem echten Gesamtergebnis. In: Praxisletter Improve Mai 3/17 (2017)
4. Reisig, W.: Understanding Petri Nets—Modeling Techniques, Analysis Methods, Case Studies. Springer (2013)
5. Inspire: Lean Office training curriculum (2018)

Chapter 5
Implementing the Lean Office

To implement successfully Lean Office, two things are not only important but necessary: knowhow and management capability. The knowhow, i.e. the "hard substance", refers to how to do it and embodies the theoretic aspect; this topic we have just been learning in Chaps. 3 and 4. In this Chap. 5 we approach the "soft substance", i.e. how to manage the implementation. Generally, people are focusing on the "hard stuff" but usually the traps to fail are linked to the "soft stuff", provided the "hard stuff" has been applied correctly. We intend with "soft stuff" the way how we behave, how we communicate, how we interact, and how we motivate employees. As we have already seen in Chap. 2, failing is exactly linked to limited or wrong knowledge and limited or bad leadership. Let us therefore switch from the technical implementation to the managerial implementation of Lean Office. Two aspects are linked to management: introducing Lean Office and managing Lean Office. Let us start first with the second aspect, which embraces the steady state of running an established lean office.

5.1 Managing Lean Office

One thing is to implement technically an office cell—another thing is to run practically an office cell or an office department. To boost not only efficiency but also efficacy, the way that an office has been managed until today, has also to change. Again, we can learn from manufacturing industry. Indeed, also the management dimension has an optimal path to follow. The management dimension should follow a Hoshin Kanri type structure of top-down target setting (e.g. [1]). Such as for shopfloor management in industrial environment, also for an office-floor management in transactional service industries, or the administrative departments of industries, a top-down target setting and a bottom-up reporting should be established. In industry, at shift handover the shopfloor teams revise together the achievements as well as issues and discuss the next steps to be done. The production managers then may report relevant information to the general management to take further appropriate decisions,

© Springer Nature Switzerland AG 2019
B. G. Rüttimann, *Transactional Lean: Preparing for the Digitalization Era*,
https://doi.org/10.1007/978-3-030-22860-6_5

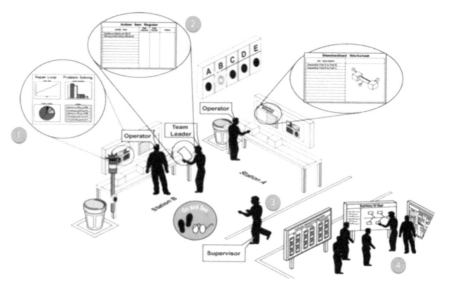

Fig. 5.1 Managing shopfloor (*Source* Escare)

if necessary. This is a very efficient and effective way to solve autonomously but coordinated management issues with respect to set targets.

Shopfloor management, i.e. the managing of a manufacturing atelier, comprises all the necessary leadership activities to deliver a certain type of performance, literally, to produce a specification-conform piece. In this context the Japanese word Gemba (meaning the location where something happens) is of essential importance because management means "going to the Gemba" (Genchi Gembutsu). In Fig. 5.1 is summarized the concept of shopfloor management. Shopfloor management is essentially based on visual management (Mieruka). These are floor markings, where e.g. containers have to be placed. But also wall signs where tools and gages have to be placed (shadow board); or Andon lights showing the status of an equipment and light boards showing to the supervisor which station has troubles and which are running smoothly; or instructions such as standard operating procedures (SOP), or relevant production information.

This leads to the

Principle of "Make the information visible and share the content with the employees".

In a nutshell: As soon as an operator detects an out-of-control situation he stops the machine turning the Andon light to red; the team leader and the supervisor will immediately hastening and trying together to detect the root cause and find a solution. The immediate stoppage of the production line is essential in the TPS because it is waste to produce defective parts; the issue has to be fixed immediately. Usually, the red light immediately signals to other up- and downstream lines to stop also production; this means it shows not only the status of the equipment but it gives compulsory indication of behavior, because it has no sense to continue to produce.

At shift end all relevant production information are summarized in key performance indicators (KPI) and compared with set targets (controlling activity), deviations are discussed and root causes identified (gap analysis), and appropriate corrective actions taken (implement improvements). Usually, for that Deming's PDCA cycle has been applied. The information is consolidated on a shopfloor management board. Balanced scorecard similar techniques are applied with information of different types such as:

– Production targets
– Quality level
– OTD performance
– or simply figures with informative character.

These visual management boards might be enriched with relevant planning information of employees (resource allocation) and acquired competencies (training level). They can be completed with ongoing Kaizen projects and their relative status of progress. The board will be the gathering point for the daily meetings (Fig. 5.2).

Shopfloor management bases heavily on the empowerment of employees. Empowerment is more than a word sleeve; if seriously and correctly implemented it embodies the engine of progress. This engine of progress, fueled by motivation, bases on the fact to considering employees not as a pure cost factor but to treat them as valuable and esteemed element of the system. This gives them the feeling of dignity to be something worth. Therefore, unchaining the potential of each employee is a leadership task and distinguishes the good from the bad manager. Managing well the shopfloor is essential (e.g. [2, 3]).

The same management approach might also be applied to office-floor. Indeed, to improve office performance in terms of efficiency but also effectiveness this industry-derived approach is compulsory. First short meeting in the morning on department level, lead by the department manager. These meeting should be scheduled every morning, not at 8:00 h when employees enter the office but e.g. at 8:45 h. This to allow employees checking first the incoming orders and e-mails in order to prioritize together at the meeting workload according to the HPFS approach. Just afterwards, a short meeting of all department leaders should be held to integrate and coordinate inter-department issues or problems, but also to discuss briefly opportunities for improvement. From these meetings may originate Kaizen or DMAIC projects.

This leads to the

Principle of "Discuss the performance daily with the employees and take appropriate actions".

Fig. 5.2 Regular shopfloor meetings can also be applied to office-floor

5.2 Office Kaizen and DMAIC

In industry, the TPS is often associated with Muda elimination and Kaizen. The Muda dimension we have already seen in Chap. 1; we have seen that to talk of Muda as epitome of TPS is fairly very limited. As we have already outlined at the beginning, the TPS is a "pull"-production theory different than usually has been applied in Western "push"-manufacturing plants. However it is true, the TPS is not only "hard stuff" such as production theory represents but also "soft stuff" how to manage continuously to improve the system. Kaizen, which is equivalent to continuous improvement, is more than a technique such as consultants are selling Kaizen, it is a way to elevate operators as important part of the system. They are not only a cost factor, but they are working at their place and they know what is best to improve the way to work. This increases their responsibility and increases their mood because they feel to be esteemed part of the system. Change Management, and how to interact with its employees, makes the difference of more or less successful companies. This needs to empower employees, i.e. giving them the knowhow and the competence to change. Training its employees is the first step to set the base of success.

Kaizen bases on Deming's PDCA cycle (Plan, Do, Check, Act) which embodies the continuous improvement culture. PDCA is derived from engineering feedback-based process control (Fig. 5.3). Although feed-forward control systems are more modern, the implementation of feedback-controlled systems constitutes already a big progress in shopfloor management. Shopfloor management is centered around Mieruka signals and Mieruka-based Kaizen boards (or shopfloor/management board however they are called) which contain information such as production achievements, employee qualifications and planning, issues and related actions as well as follow-ups, as we have seen in the previous section. These boards serve at coordination meetings to align the operators with the necessary information to do best their work. Encountered issues are discussed and, if possible, immediately solved at these occasions using PDCA approach. Usually, these meetings take place in the shopfloor area at shift begin to align the new team with open issues of the former shift.

Generally, manufacturing issues may be of two different types: small ones with the Andon light of the cell or transfer line switching from green to orange signaling

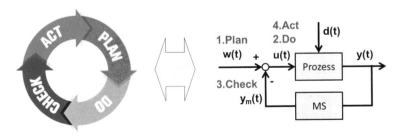

Fig. 5.3 Deming's PDCA continuous improvement cycle bases on process control theory

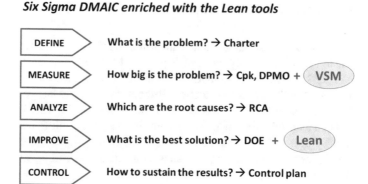

Fig. 5.4 LSS DMAIC—a proven problem solving approach

performance loss on the one hand and on the other hand big issues where the Andon light switches to red and the line comes to standstill. In the first case, the FIFO WIP-capped transfer line allows to continue downstream production for further some minutes while finding a quick solution to the issue. In the case of a big failure however everybody stops working and helps to solve the problem. The problem can be compensated without having identified the root cause or the Kaizen team fixes definitely the problem utilizing the Ishikawa diagram to find the real root causes. If problems reveal to be of extensive character, even a Lean Six Sigma (LSS) DMAIC team (Define, Measure, Analyze, Improve, Control) may be put together to fix the problems definitely [4, 5], after having found in advance a palliative solution. Indeed, solving also temporarily the problem is essential because a DMAIC project may also last up to three months. In Fig. 5.4 are shown the DMAIC phases and its core contents and deliverables.

Now, the same concepts can be transferred to the office environment. OPEX initiatives in Europe have already largely been implemented using the LSS DMAIC problem solving approach. The initial success has been followed by declining results. The reason is linked to the method. As we have seen in Chap. 2 LSS DMAIC is apt to solve problems but not to change a culture. The DMAIC approach is always good to be used for complex problem solving and is heavily recommended to be used also in an office environment. However as already described at the beginning, to change the culture of working we have to implement a Lean transformation approach and apply the here presented comprehensive Lean tools system.

This leads to the

Principle of "Empower the employees and install a continuous improvement culture".

5.3 Deploying Lean Office in the Company

With the term operational excellence (OPEX) is intended a management philosophy to strive according to the continuous improvement approach with unrelenting passion to always new performance heights. Whether it has been implemented by applying a LSS DMAIC, Lean transformation, or Lean introduction, or combined approaches, generally had contingent reasons. Fact is, whereas this approach has been widely implemented in manufacturing industries, service industries have initiated but hardly completed the transformation.

It has to be said right from begin: To industrialize the office world in order to boost office-floor productivity is a real challenge. The challenge is not to implement the here presented theoretic concepts, the challenge will be to overcome the resistance of office white-collar employees. This is because since ever office employees generally benefitted from not strictly supervised work control; being now time-controlled will for sure influencing their mood. However, why should white-collar employees be treated differently from blue-collar workers? Indeed, those employees, which recognize the need and benefit, will contribute to the future of their company and secure their work places. These employees have to be appointed Lean office change agents for their departments; the employees showing resistance, it might sound hard, have to be dismissed. A company cannot allow that the future survival of the company is endangered by non-cooperative employees.

To deploy the industrialization of transactional office activities, it is advisable to be accompanied by a LSS MBB or real Lean expert. Not only mastering the tools but also the knowledge behind of what Toyota's thinking consist of is mandatory. Each small setback when trying to transform an established way to work will be taken as pretext to say it does not work. Change management (CM) will be of crucial importance. It is recommendable to establish a multi generation plan (MGP) when implementing Lean office.

- First generation: start with a pilot implementing an office cell and draw lessons learned.
- Second generation: enlarge to the whole department or business unit by applying lessons learned and define best practices.
- Third generation: rollout to the whole enterprise group making extensive use of best practice.

The rollout should be implemented with dedicated department leaders. Before rolling out Lean office, the standards of Mieruka have to be set. Visual management is very important because it facilitates the working. Indeed, on the shopfloor WIP is immediately visible; on the office-floor WIP is hidden in computers or desks. It is therefore necessary to make WIP visible and show visually the actual workload. Visual management is not limited to the colors of folders but could "in extremis" also be extended to the use of Andon lights or similar devices such as simple flags or electronically managed lights in virtual office cells. Practical aspects are treated in the Chap. 4.

For further information, consult change management experts. To master the new digitalization era, lean experts as well as IT specialist have to work closely together. In the next section, we will cover briefly the business excellence (BEX) dimension, i.e. integrating strategy into OPEX. Indeed, in the new digitalization era BEX will not substitute OPEX but OPEX will evolve into BEX because digitalization will allow new business models.

5.4 Preparing for Digitalization

If some managers think, that the new digitalization era will solve all their problems, such as "deus ex machina" in ancient Greek theatre, these adult managers may still believe in Santa Claus. To the contrary, digitalization is a new challenge to master. It is forecasted, that office digitalization thanks to robotic process automation (RPA) as well as artificial intelligence (AI) will not only increase productivity by decreasing white-collar employees, but will allow completely new business models (see Sect. 6.4). In industry, extreme digitalization by fully interconnecting all economic agents is named the 4th industrial revolution (see Sect. 6.3). However, these changes have not the same disruptive character such as pure technical innovations (e.g. electronic valve vs transistor, or carburetor vs. fuel injection) where the user is not concerned. The new digitalization era will take time because digitalization is a systemic challenge—the digital transformation will take time but it will take place for sure. Therefore, companies have time to prepare. The new system of digital revolution can simply be described as being composed of

– IT standards
– Telecommunication infrastructure
– Integrated data base
– Intelligent software
– Company knowhow
– Human behavior.

Such as the digital technology is not yet fully available and has to be developed, the new competence is by far not yet available in all companies and has also to be formed. Further, the most resistive component, i.e. the human beings, will have to learn and to adapt to the changing way to do business or even to the changed way to live. However, not embarking the emerging trend will for sure compromise the future of the companies. However, how to approach this new digitalization challenge? Will digitalization replace lean thinking? Will it be possible to sublimate, i.e. becoming digital without having been lean? Maybe—maybe rather not! Firstly, it is true, digitalization might set new competition rules; however, secondly, digitalization is also a vehicle to implement a solution. Due to the fact, that digitalization will lingering finding their way, we have to survive the period until the company has completed the digital transformation and is ready for the future. Manufacturing industry has nearly completed the lean transformation and is ready for the "Industry 4.0 Challenge";

however, service industry is yet still far away. As long as processes govern the transformation of inputs into output, the Lean theory will have its place. If a different and hypothetic new "post-process non-processic era" will materialize, nobody knows. Therefore, Lean office will remain a first, basic prerequisite for digitalization.

It might sound silly, but simple paper-based cells are the first step to initiate to face the digitalization era. It is the prerequisite for the procedural electronic cell. Indeed, if the workflow is not precisely defined and the work-interaction has not run-in well, to digitalize the process in classical manner will be difficult. If we define a logic office transformation-approach, a meta-process to become "digital", it will look like this:

– first, install a paper-based work-cell to run-in the process (increase productivity)
– second, transform the cell into an electronic file-based processing (eliminate paper)
– third, evolute the process into an AI-based application (change paradigm).

It has to be precised, the new digitalization has not only the taste to be a disruptive evolution, digitalization needs also prerequisites to be implemented. One basic prerequisite is lean-based processes. Indeed, the first step will increase productivity and speed; becoming lean means to eliminate obvious Muda such as waiting and non-adding value tasks such as manual data input. To do so, SFE stands at the base as well as applying all the enounced principles in this book. The first step might have severe consequence on white-collar employment because it reduces a lot of paid waste in form of waiting, i.e. doing nothing. This step is the most important step, which also does not imply much spending. Indeed, the process have to become performant increasing productivity, but neither the provided service nor the service level have to be reduced by dismissing employees! Unfortunately, exactly this has been observable in recent restructuring initiatives of credit institutes. The second step of our meta-process of transformation prepares the data structure to be consistent and not redundant. This topic is not new, but it has to be solved for efficient electronic office cell implementation. This might lead to expensive change of the IT system. Not only the vulnerability of the IT system but also the development degree of IT systems are to be desired and costing a lot of money. At the end of this step, the electronic file-based cell might also evolve into a virtual office cell, where the concomitant physical presence of employees is not anymore required. The third step consists of the change of present paradigm—the new technology will allow new business models. However, this is another topic not covered in this book.

The just made reflections show, that lean is not a choice—to be lean is mandatory. This leads to the

Principle of "Lean-up processes to prepare for the digital transformation".

Indeed, during the last two decades, the increased communication possibilities favored by internet have fueled globalization enabling a rapid increase in world trade as well as financial transactions with positive and negative consequences (e.g. [6, 7]). The globalization challenge had imposed to become lean (Fig. 5.5).

Not being lean has been meaning to suffer considerable competitive disadvantages in a highly globalized competitive context regarding productivity and response

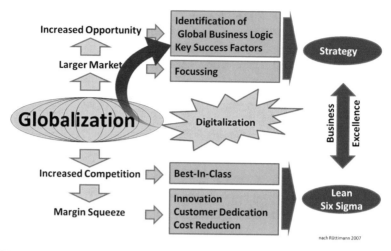

Fig. 5.5 Digitalization opening a new globalization era with new industry logic (adapted from [8])

rapidity. Now, increased digitalization through big data and internet of things (IOT) is changing again industry logic opening the possibility for new business models. The combination of Lean (as well as Six Sigma quality) with new strategies based on digital business models will evolve OPEX into BEX. OPEX will only cover the basic requirements to become best-in-class. The countdown has already beginning.

References and Selected Readings

1. Akao, Y.: Hoshin Kanri: Policy Deployment for Successful TQM. Taylor & Francis Limited (2017)
2. Suzaki, K.: The New Shopfloor Management—Empowering People for Continuous Improvement. Free Press (2010)
3. Scherer, E.: Shopfloor Control—A Systems Perspective: From Deterministic Models towards Agile Operations Management. Springer (2012)
4. George, M.: Lean Six Sigma: Combining Six Sigma Quality with Lean Production Speed. McGraw-Hill
5. George, M.: Lean Six Sigma for Service. McGraw-Hill (2003)
6. Rüttimann, B.G.: Long-term international trade analysis measuring spatial extension of globalization: kuznets or hysteresis paradigm? In: Proceedings in Business and Economics (ICOAE 2018). Springer Nature Switzerland (2018)
7. Rüttimann B.G.: Modeling financial type 2b globalization and its effects on trade, investments and unemployment. IJCEE **6**(2) (2016)
8. Rüttimann B.G.: Introduction to Lean Manufacturing and Six Sigma Quality Management. ETH Zürich, Lecturing Notes (2017)

Chapter 6
Some Additional Aphorism

In the former chapters, we have learned how to boost productivity of transactional processes by implementing the Lean Office transformation. In this final chapter, we will enter a brief discourse linked to current lean office topics. We show exemplarily how to shorten timeline of the accounting closing process and how our Western mindset of accounting rules does not support Lean thinking. Conclusions will lead to additional Lean Office Principles. An outlook on Industry 4.0 applied to office and how business models may evolve will close the overview of the necessary lean office transformation in order to anticipating the latent but imminent office revolution.

6.1 Practical Example: The Lean Closing Process in Accounting[1]

For most companies, closing of the books is a painful and time-consuming process, which is required by law and maybe investors but does not add value as such. Therefore, the customer does not particularly care and moreover is not willing to pay for this closing process; it belongs typically to the category of "process required" NVA activities. This is the reason why it has not been on the top of the OPEX agenda.

Some companies have different procedures and contents for monthly, quarterly or yearly closing. A typical month- or year-end closing involves the following activities and timelines, where the below alphanumeric figure means the month-end date plus the working days to perform typical closing activities:

Month-end	D
IC plus Netting Supplier Input	D + 3
IC Counterpart Input	D + 4
IC Netting Input and Reconciliation	D + 5

---

[1] This section is the contribution of Urs P. Fischer, Former CFO of Alusuisse and A-L Group, now founder and president of Leansolution, Switzerland.

© Springer Nature Switzerland AG 2019
B. G. Rüttimann, *Transactional Lean: Preparing for the Digitalization Era*,
https://doi.org/10.1007/978-3-030-22860-6_6

Pre-Netting Run at Group	$D + 6$
Final Netting Run by Group	$D + 7$
Submission P&L and Balance Sheet incl. Cashflow	$D + 7$
IC Settlement of Net Positions	$D + 10$
Closing and Consolidation	$D + 10$
Reporting	$D + 11$

This means, that the final report is ready after eleven days of month-end, or in other words, the necessary time with all required activities to create the report takes eleven days. Alone the time to make the IC reconciliation takes five days (most of it waiting time). Why should we care so much about the reconciliation of intercompany (IC) positions? If a group internal supplier posts revenues without its counterparty posting the related charges, we would create an artificial profit for the group. In the following month, the group internal customer would then post the charges, which would subsequently create an artificial loss for the group. Therefore, after two months, the "profit" and "loss" would cancel out and we are again fine, but the local and generally accepted accounting principles do not allow this asymmetric treatment of income and cost. Therefore, the reconciliation of intercompany credits and debits is a prerequisite for the closing. At the same time, it can provide the basis for a settlement and netting of cash payments in the group. Furthermore, foreign exchange positions can be netted as well and, if the policy requires, remaining exposures hedged.

The intercompany reconciliation is time consuming for various reasons. Firstly, there may be delays between the issuance of an invoice in the supplying company and its reception in the receiving company. Also, company internal processes (e.g. mandatory sign-off by purchasing staff, goods receipt department or higher management levels) may be slow and lead to further latencies and result in larger imbalances. For that, certain solutions have been outlined in the former chapters. However, the contingent situation needs to be analyzed case by case. Secondly, in a global group, working time across the world do not match. When employees start their business day in the US their colleagues in Asia are already at home. So, we need typically a few days to sort out mismatches between group internal creditors and debtors.

Now let us have a look at this problem from the lean point of view. If we wait until the month end to start this reconciliation, we pile up an "inventory" of non-resolved issues during the month, that all must be dealt with a few days after month end. Exactly in the period, where accounting staff is already busy for other reasons (e.g. their local closing process). Most of these issues could have been clarified already during the month making better use of existing resources.

The timeline of the revised closing process might look as follows, where i denotes the monthly days (i = 30, 29, 28,... 3, 2, 1):

IC plus Netting Supplier Input (continuously)	$D - i$
IC Counterpart Input (continuously)	$D - i$
IC Reconciliation	$D - i$
IC Netting	D
Pre-netting Run at Group	$D + 1$
Final Netting Run by Group	$D + 2$

The potential future close-calendar

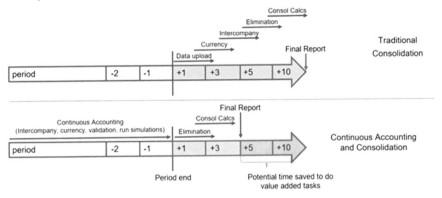

Fig. 6.1 Comparison of traditional versus improved monthly closing timeline [1]

Submission P&L and Balance Sheet incl. Cashflow	D + 2
IC Settlement of Net Positions	D + 3
Closing and Consolidation	D + 3

Therefore, if we move to permanent clarification during the month, we are much closer to "one-piece flow" in accounting and can considerably shorten the closing process. This approach is a key element of continuous accounting and consolidation as summarized in Fig. 6.1.

This is primarily a different accounting and consolidation process and does not necessarily require sophisticated and harmonized IT across the group. However, modern integrated software systems do of course simplify this change as well. Apart from shifting reconciliations from monthly to continuous, we can further support by tools that automatically recognize and help to match intercompany positions even if they are in different currencies, keep open intercompany positions in an environment, that can be accessed by all parties involved (e.g. cloud), increase visibility by tracking the progress of required approvals, get access to further documents (orders, contracts) that are related to the invoices, ease substitution of individuals, etc.

Shortening the closing period does not provide value as such but getting beyond allows to focus resources on more important value-added tasks (e.g. analyze markets and competitor moves, work on investments and performance improvement projects, etc.).

Anticipating tasks before their partial results are needed seems not to comply with the lean JIT logic. However, this is only a superficial conclusion; indeed, the IC reconciliation of a single transaction is only completed when the corresponding bookings are terminated. This equals a SFE at the occurrence whereas the waiting for all IC bookings and clearance would be a batch operated mode. However, the anticipation of work may also have another reason. If the necessary capacity, i.e. the required workforce, cannot be increased rapidly at the closing moment, the workload of the single employee has to be leveled by anticipation.

This leads to additional principles directly derived from the Principle of "If possible, strive for sub-routine or even end-to-end process completion" such as, the

Principle of "Take a transaction only one time in the hand and complete it whenever and as soon as possible"

and the

Principle of "Level temporal workforce constraints by anticipating activities".

This principle also shows, that the implementation of an artificial backward pulling SFE makes no sense (see Sect. 3.8). Instead, the order is pushed in a controlled manner through the organization by CONWIP technique. It has to be clarified, as already stated in Chap. 2, due to the different characteristics of office transactions but also how the service is provided (impossibility of "chaku chaku") the way to work in offices differ from manufacturing, leading to specific office-dedicated lean principles not applied in manufacturing and even contradicting the JIT mantra.

6.2   Critical Thought: Traditional Accounting Rewards Muda[2]

To lean-up operations and support functions is key to survive in today's competitive environment. However, not only the production process but also the management and controlling process has to comply to synergic requirements in order to support the Lean philosophy. Whereas JIT production as well as Hoshin Kanri-based management supports Lean, the traditional accounting system does not yet support Lean.

Accounting would be very easy if all the products produced would be immediately sold against cash. Let's take the example of a mobile chicken grill owner. He buys in the morning chickens, grills and sells them during the day to customers that pay in cash. Once he has sold all his chickens, he shuts down his business overnight and can now calculate his profit. If we ignore for simplification the amortization of equipment (fixed cost) and the cost of energy (variable cost), his profit is equal to the cash in hand at the end of the day less the cash he had before buying chickens. Profit is in this business equal to cash flow as there are no raw, semi-finished or finished products inventories to be considered.

Few businesses are like that. Most businesses will not immediately sell products against cash. The production of goods may last days, weeks or months. Or we may for technical reasons have to produce batches that are larger than the daily consumption. Then, calculating the profit gets much trickier, as we must value intermediate and finished products. Indeed, accounting rules allow to value intermediate and finished products including large parts of production cost.

---

[2]This section is the contribution of Urs P. Fischer, Former CFO of Alusuisse and A-L Group, now founder and president of Leansolution, Switzerland.

P&L with Stock Production		P&L without Stock Production	
Sales	600	Sales	600
Stock change	100	Stock change	0
Total Revenues	**700**	**Total Revenues**	**600**
Material	-250	Material	-200
Energy	-100	Energy	-100
Labor	-250	Labor	-250
Depreciation	-100	Depreciation	-100
Profit	**0**	**Profit**	**-50**

Cashflow with Stock Production		Cashflow without Stock Production	
Profit	0	Profit	-50
Depreciation	100	Depreciation	100
Cashflow	**100**	**Cashflow**	**50**
Change in Inventories	-100	Change in Inventories	0
Change in Cash	**0**	**Change in Cash**	**50**

Fig. 6.2 Effects of overproduction (didactic simplified example)

Such valuation is often made using standard cost. Standard Costing is an accounting technique in which standard product costs are calculated for every part, sub-assembly, and finished product, i.e. every step in production. The word "Standard" has a meaning like expected long-term average or normalized. The inventory value is then computed by multiplying the number of products in each intermediate or final stage with the according standard cost. Any increase in inventory value during a period is considered as income, decreases as cost and build a complementary part of the income statement together with revenues from sales and personnel and all other cost. As it is not yet sold, it is posted in the balance sheet as an asset.

Inventory valuation can also be based on actual cost, but this procedure requires that detailed "actual" information about all cost categories, material consumption, production times, and scrap are gathered. Typically, a sophisticated costing system is complex to set up and requires millions of wasteful and time-consuming entries of the "actual" data. I am using the word "waste" in the Lean context as "any activity that does not add value in the eyes of the customer."

The inclusion of current production cost in the inventory valuation creates an incentive for management for overproduction and inventory building if market demand is low. This is explained in Fig. 6.2. On the left side we produce stock valued 100, assuming 50 material cost and 100 for production cost, and are able to break-even on profit level. On the right side, we do not produce on stock and record a loss of 50. Obviously, management prefers in this situation to overproduce if it is measured by profit.

Fig. 6.3 Different value system leads to different control system

If it is measured by cash flow, not producing on stock is more favorable. Why is this? If we produce on stock, we have to replace the used raw material, because if we don't, we may not be able to respond to the next customer order. By moving from raw material to finished goods we have lost the flexibility of the raw material availability. If the next customer order does not exactly match our product hold on stock, we have to use again raw material and produce it as required by the customer. So, on the left side we spend 50 to replace used-up raw material whereas on the right we can keep the 50 cash and pay back debts.

This implies that the management control system is in line with the management value system that is given by their predominant production system used. More general, traditional and lean performance measurement systems have very different focus and this fact results in completely different incentives and drivers as shown in Fig. 6.3.

Conclusion: If we do not want to reward production of Muda we should use an accounting and performance measurement system that is incentivizing lean. Such a system should be based on cash flow and capital employed and not book profit as in traditional accounting. For further information consult e.g. [2].

6.3 Paradigm Change: Office and Industry 4.0

The service industry is not based on physical products, but is based on transactions. The term transaction has been coined in the era of electronic data processing (EDP). Today we call it not anymore EDP but information technology (IT). The terms big data and artificial intelligence (AI) are widely used, although increased digitalization might be a better word; anyway, all that is today summarized under the buzzword of Industry 4.0 revolution. It does not come by chance, but service industry without IT is not imaginable. Today the office IT is the central backbone of any business and the fast advancing digitalization will also influence the office environment.

The catchword Industry 4.0 depicts how business will be in future. With the internet of things (IOT) products and machines become identifiable and will be

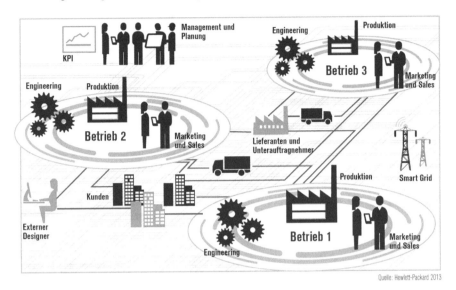

Quelle: Hewlett-Packard 2013

Fig. 6.4 Horizontal networking within a CPS (from [3])

able to communicate between them. Delayed trucks will interfere with production scheduling, additive manufacturing technology with 3D printers will allow to fabricating customized parts on demand, and products will find themselves their optimal production sequence. The concept bases on a cyber-physical system (CPS) which interconnects horizontally different agents of our socio-economic world (Fig. 6.4). Supplier and customer will be close in touch.

The Industry 4.0 concepts will also influence heavily shopfloor (vertical dimension). It is not the intention to develop here a critical analysis on Industry 4.0 but let us make a small excurse to the shopfloor scheduling. Today's "fixed-wired" production schemes are intended to be substituted by "flexible logic" production schemes allowing even batch-size one, i.e. a one-off [3]. The scheduling program is even intended to be optimized interactively by the smart objects (e.g. products and machines) interconnected within a cyber-physical production system (CPPS). How the German Acatech group is envisaging, this scenario is shown in Fig. 6.5 (by Hewlett-Packard [3]). How realistic this will be we leave here apart; indeed, the concept of Pareto optimality will be very difficult to be observed and Muda will materialize despite AI-technology [4]. In any case, certain things sound for the time being rather impossible or economically not viable (e.g. confront not used capacity of several stations in Fig. 6.5 [3]), however, other things will become soon reality. Nevertheless, Industry 4.0 will change production but above all, it will allow to doing business differently.

If we coin the word Office 4.0 then we could imagine a similar scenario. Apps will facilitate the work of administrative employees, smart watches with sensors will reveal our mood and interfere with room ambiance changing temperature, illumination, and music, leading to a "cognitive environment". If also an intelligent front desk

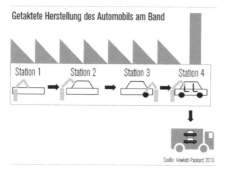

Heute

Getaktete Herstellung des Automobils am Band

Station 1   Station 2   Station 3   Station 4

Quelle: Hewlett-Packard 2013

In der Automobilproduktion existieren heute zwangs-verkettete Produktionsstraßen; die Umrüstung für neue Produktvarianten ist aufwendig. Die durch IT-Lösungen unterstützten Produktionsleitsysteme (Manufacturing Execution Services, MES) sind meistens entsprechend der Hardware der Produktionsstraße mit genau defi-niertem Funktionsumfang konzipiert und daher statisch. Die Arbeit der Beschäftigten wird ebenso durch die Funktionalität der Produktionsstraße definiert und ist in der Regel monoton. Eine Berücksichtigung individuel-ler Kundenwünsche, wie etwa der Einbau eines Ele-mentes aus einer anderen Produktgruppe des gleichen Unternehmens – beispielsweise Porsche-Sitze in einen Volkswagen –, ist nicht möglich.

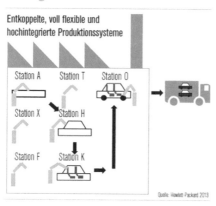

Morgen

Entkoppelte, voll flexible und hochintegrierte Produktionssysteme

Station A   Station T   Station O

Station X   Station H

Station F   Station K

Quelle: Hewlett-Packard 2013

In Industrie 4.0 entstehen dynamische Produktionsstra-ßen. Das Fahrzeug fährt darin als Smart Product auto-nom durch CPS-fähige Prozessmodule des Montage-raums. Durch die dynamische Umrüstung der Produktionsstraßen wird ein Variantenmix in der Aus-stattung möglich; einzelne Varianten können zeitunab-hängig von der vorgegebenen zentralen Taktung und reaktiv auf Logistikaspekte (Engpässe etc.) vorgenom-men werden. Die IT-Lösung für das Produktionsleitsys-tem ist nun eine zentrale Komponente– von der Konst-ruktion über die Montage bis zur Inbetriebnahme. Das Leitsystembesteht aus *Apps* basierend auf einem Ma-nufacturing Operation System (MOS) und einer föde-rativen IKT-Plattform. Apps, MOS und die Plattform re-alisieren nun das neue flexible MES. Durch die neue Dynamik kann der Einbau individueller Elemente (Por-sche-Sitz) problemlos integriert werden.

Fig. 6.5 How Industry 4.0 shopfloor is intended by the acatech group/Hewlett-Packard (from [3])

computer will have conversation with potential customers is only a question of time according to AI progresses; if that will be well accepted by clients we have to test— first applications are appearing. However, if the only differentiating ace that service companies have, i.e. the interaction with the direct client, will be left to machines will be a very intelligent move, I doubt heavily—but we will see. In any case, employees and customers become part of the IOT. If that is desirable or not we will not discuss here and leave it apart.

Beside the amenities of an intelligent cognitive environment, the way how office work will evolve is not yet clear and how much Industry 4.0 will interfere with white collar occupation is not the topic of this section. However, the way how employees work in an office will for sure change. Since more than 30 years, the paperless office has been predicted, but the paperless office is advancing only slowly. The technology to implement already exists. This resistance is not only linked to the human beings preferring paper (paper books are still printed although electronic versions are available) which might be linked to generational behavior, but also to the fact that often orders are still paper-based and the incoming papers have first to be scanned or have to be copied manually leading potentially to mistakes. As soon as IOT and acceptance of digital interfaces evolve, the prerequisites are given to let the paperless office materialize. However, if machines will substitute office-floor employees such as robots are substituting shopfloor operators we will see. We have always to keep in mind, although it might sound heretic: an accountant is not hired primarily to enter commercial transaction data into computer but to take decisions. In the same sense, the order executing operator at the assembly line will evolve and become a problem solving operator. The stochastic character of orders with variable content will take a little bit more time to be solved with AI algorithms. During this time, it is recommended to eliminate the huge waste existing in transactions and to boost productivity in the offices according to the here presented concepts. Industry or better Office 4.0 will change the way we work, if initially not focused on the front customer interaction allowing new business models, at least monitoring the status of transaction advancement and facilitating data analysis. However, remember, basic production laws governing performance, still remain valid and, therefore, also Lean theory remains valid. Instead of waiting for the promised land of Office 4.0, it is heavily recommended to undertake the Lean transformation of the office environment, which houses a lot of Muda leading immediately to an enormous tangible effect assuring the future survival of the company. Lean as the prerequisite for the digitalization - otherwise, we are digitalizing Muda.

Indeed, this leads to the last and ultimate

'Guiding Principle' of "Before trying to fly we should learn to walk because we are still stumbling".

6.4 Outlook: New Digitalization-Based Business Models in Insurance[3]

For several decades, companies have used lean principles and tools to reduce operational complexity and to improve productivity. The Lean Office approach provides the foundation for operational excellence by standardizing processes, establishing a culture of continuous improvement, and empowering employees on the shop floor. However, Lean Office is limited in its opportunities and will not be sufficient to survive in the future.

Insurers need to use the potential of digital technologies to persist on the market. Consolidated and simplified processes across the entire company combined with an

---

[3]This section is the contribution of Antonio Gallicchio, Master Black Belt and OPEX director at Generali Insurance Company, Switzerland.

effective digital strategy are therefore essential to manage the digital transformation and to ensure long-term success [5]. Innovations are indispensable for both to stay ahead of legacy competitors and disruption of established business models by ambitious digital start-ups.

Aspects to consider for the digital transformation are:

- customer expectations
- business models
- machine learning and advanced analytics
- skill shift.

Let us have a brief look at each aspects.

Customer expectation: With digitalization and simplified access to new technologies, customer expectations and needs are changing and increasing. For that, insurers need to rethink and transform the customer experience and journey. To enhance the relationship with customers the interaction with customers is becoming more important and requires new products, tools and services [5].

Business models: Chia Tek Yew describes the need of new business model very precisely: "Disruptive innovation is both the most meaningful and most challenging to achieve. Instead of looking at specific technologies, disruptive innovation is about creating a vision for a new insurance business model, supported by technology. Today's business models will not survive in tomorrow's digital landscape - insurers must evolve to survive. To achieve the vision of an insurer being at the front end of a customer's daily transactions, the way that an insurer operates needs to be reconsidered from the ground up" [6]. The paper from McKinsey (Digital insurance in 2018: Driving real impact with digital and analytics) elucidate the changes in the insurance sector and emphasizes the need of redefining boundaries across industries. "Traditional industry borders will fall away, the future of insurance stands to be influenced by platforms and ecosystems" [7].

Ecosystems are dynamic associations of various organisms of different categories, which interact with each other as functional unit. Such interactions and collaborations with different industries become more important due to the chance of generating new opportunities. Companies like Amazon, Apple or Microsoft are examples of player of such ecosystems. They create new customer experiences by integrating linked sets of services to allow customer to fulfil different needs from one hand [7]. According to McKinsey, 12 ecosystems exist by 2025 that revolve around human and organizational needs (Fig. 6.6) [7].

Machine learning and advanced analytics: With new technologies, the volume of data will increase significantly. The more connected we get the more data is generated. Big Data can help insurer to gain insight of their customers. This helps managing risks in an advanced manner, generating new business opportunities and making significant cost savings [5]. "Where operations rely on extensive processing of 'personal data' they can give rise to data protection and privacy concerns in many countries, particularly under the new European General Data Protection Regulation. For many insurance players navigating the complex landscape of data protection,

New ecosystems are likely to emerge in place of many traditional industries by 2025.

Ecosystem illustration, estimated total sales in 2025,[1] $ trillion

[1] Circle sizes show approximate revenue pool sizes. Additional ecosystems are expected to emerge in addition to those depicted; not all industries or subcategories are shown.

Source: IHS World Industry Service; Panorama by McKinsey; McKinsey analysis

Fig. 6.6 Ecosystems in 2015 (*Source* McKinsey)

privacy and cyber law and regulation will be key to unlocking the full power of Big Data, advanced analytics and machine learning" [5].

Skill shift: Digital Transformation without changes in culture, in the adaption of the way of working and without a shift of people skills is not possible. Human resources are facing new challenges to restructure the existing staff and to find solutions to attract the right "digital" talents. The latter is one of the key competitive advantages that companies have [5].

To conclude, operational excellence is a "conditio sine qua non", however, which itself is not enough to survive tomorrow's challenge, not being the final solution any more to create long-term value in today's changing environment. It is merely the base for radical simplification of business process. This is the start of a digital journey which includes more than only the use of new technologies even though the latter drives the whole digital transformation.

References and Selected Readings

1. Milne, E.: White paper: Continuous Accounting Drives Strategic Transformation. SAP AG (2017)
2. Lean accounting: management for the lean organization www.maskell.com
3. Arbeitskreis Industrie 4.0, April 2013, Umsetzungsempfehlungen_Industrie4_0.pdf, closing report published on www.plattform-i4.0.de, acatech
4. Rüttimann, B.G., Stöckli, M.T.: Lean and industry 4.0—twins, partners, or contenders? A due clarification regarding the supposed clash of two production systems. JSSM **9** (2016)
5. DLA Piper: Digital Transformation in the Insurance Sector. White paper (2016)
6. Chia Tek Yew, partner and head of financial services advisory, KPMG Singapore
7. McKinsey: Digital Insurance in 2018: Driving real impact with digital analytics (2018)

Epilogue

This book is a first attempt to describe a TPS-derived industrialization of office environment in order to increase transactional process performance in view of the upcoming new digitalization era. It is an alternative to the present prevailing Muda-centric identification approaches mainly based on swim-lane analysis, which are justified at the beginning of an OPEX initiative, but which show very soon its limitations. However, as we have seen, the here presented approach has a more holistic scope using an industrial-based paradigmatic approach to the transformation of input into output; it is a profound change to the way how employees may work today in the office and a first step towards digital office cells. Indeed, we can make an allegoric comparison: the present office organization approximately corresponds to the Muda-poisoned Western B&Q push manufacturing mode with low performance, whereas the here presented office cell organization or lean SFE corresponds to the Toyota derived demand pull SPF, which shows a higher performance in terms of throughput and speed.

Nevertheless, we have to be aware, the office is not the shopfloor, which has different characteristics with regard to transactions; the TPS tools have been developed for a JIT manufacturing environment. The transfer is only in contingent situation feasible and has to be reinterpreted. This reinterpretation is suitable for a restricted type of transactions. In this book were outlined the reinterpretation of the concepts suitable to structured processes for an office industrialization to boost productivity. Further, in industry the production pace are seconds and minutes with reproducible content, whereas in the office the pace are minutes and hours with a non-deterministic content. Nevertheless, this does not mean that the industry concepts are not valid for the office, but they have to be reinterpreted and adapted—one could say "…ad augusta per angusta" or more appropriate "…per aspera ad astra". The stake is worth to try it.

© Springer Nature Switzerland AG 2019   113
B. G. Rüttimann, *Transactional Lean: Preparing for the Digitalization Era*,
https://doi.org/10.1007/978-3-030-22860-6

The challenge to industrialize and digitalize the office environment to face the new digitalization bet is big and the chance to fail is even bigger. Therefore, I want to enounce the last principle, the

Principle of "Do not surrender and live the Kaizen philosophy";

a setback is a new chance to retry with increased experience. If you have any problems, mail to brunoruettimann@bluewin.ch

Overview of the 25 Lean Office Principles

In the following, we summarize the list of practical Lean Office principles and the relative chapter where they have been enounced. Please notice, you are looking perhaps in vain for the enouncement of one of the key Lean beliefs for shopfloor such as "going to the Gemba". Indeed, because Muda is not always immediately visible on office-floor (with e.g. the exception of going with documents to the copier as well as the overfilled paper-based inbox), I preferred not to list it as principle. However, this is not an excuse not to show presence. This not enounced principle has been substituted on office-floor by the first enounced principle: make extensive use of graphical process description. This is another evidence for the difference between Lean shopfloor and Lean office-floor.

Chapter 3: The Office-Adapted Lean Tool System

> *Principle of "Make extensive use of graphical swim-lane-type of process flow description to understand service performance generation"*
> *Principle of "Concentrate the main attention always on the bottleneck because it limits directly throughput"*
> *Principle of "Strive for uniform layout of the office desk workplace"*
> *Principle of "Observe the repeatability and reproducibility of a task or a routine for every employee"*
> *Principle of "Strive-for-team-efficiency in order to maximize productivity"*
> *Principle of "Measure your performance to be able to improve performance"*
> *Principle of "Assure VOC conformity by first time right"*
> *Principle of "Always complete a task before interrupting the job"*
> *Principle of "If possible, strive for sub-routine or even end-to-end process completion"*
> *Principle of "Pull if you cannot flow"*
> *Principle of "SFE forwarding instead of batch forwarding to reduce PLT"*
> *Principle of "Output triggers the release of new orders or, input equals output will stabilize PLT"*

© Springer Nature Switzerland AG 2019
B. G. Rüttimann, *Transactional Lean: Preparing for the Digitalization Era*,
https://doi.org/10.1007/978-3-030-22860-6

Principle of "Keep employees always busy with VA-activities, i.e. eliminate NVA-activities"
Principle of "Install priority-pull or HPFS scheduling in order to increase effectiveness"
Principle of "Put-efficacy-over-efficiency".

Chapter 4: Office Cell Design

Principle of "Inter-department office cells will speed-up PLT of end-to-end transactions"
Principle of "Advantage of integrated job execution over tayloristic-split job execution"
Principle of "Maximize parallelization of work as much as possible"
Principle of "Lean-up processes to prepare for the digital transformation".

Chapter 5: Implementing the Lean Office

Principle of "Make the information visible and share the content with the employees"
Principle of "Discuss the performance daily with the employees and take appropriate actions"
Principle of "Empower the employees and install a continuous improvement culture".

Chapter 6: Some Additional Aphorism/Epilogue

Principle of "Take a transaction only one time in the hand and complete it whenever and as soon as possible"
Principle of "Level temporal workforce constraints by anticipating activities"
'Guiding Principle' of "Before trying to fly we should learn to walk because we are still stumbling"
Principle of "Do not surrender and live the Kaizen philosophy".